Year A Teacher Guide

Science Chunks - Year A

First Edition 2021

ISBN: 978-1-953490-09-4

Email: support@elementalscience.com

Copyright Policy

Year A Overview

Welcome to Year A of the Science Chunks series! The seven units that make up this year will cover plants, biomes, the solar system, atoms, molecules, light, and sound, plus a short unit on Gregor Mendel. You will learn about these subjects by:

- **Reading** the assigned encyclopedia pages with your students to learn about the lesson's topic.
- **Writing** down what the students have learned in a way that is appropriate for their skills using either the lapbook templates or the notebook sheets.
- **Doing** one or more of the lesson's coordinating STEAM activities.

What do we need?

In addition to the lesson plans found in the seven units included in Year A, you will need the following materials:

1. **Required Books** – See a complete list on p. 8 of this guide or visit the Year A resource page for links to the books:
   - https://elementalscience.com/blogs/resources/sc-a
2. **Student materials** – Unlike the single Science Chunks units, this printed Year A guide does not inlcude the student materials. You will need to purchase one (or both) of the following:
   - *Science Chunks Year A Student Lapbook Templates (LT)*
   - *Science Chunks Year A Student Notebook (NB)*
3. **Supplies** – You will need various supplies for the lapbook or the notebook, plus the coordinating hands-on science activities. You can find a full list of what you need for the year on pp. 11-12.

How do we organize the year?

Generally, we recommend completing one lesson per week. Please feel free to organize the order of the units in a way that fits your goals. That said, here are two suggestions:

Option 1 (By Discipline)

- [ ] Plants Unit (p. 13)
- [ ] Mendel Unit (p. 27)
- [ ] Major Biomes Unit (p. 31)
- [ ] Solar System Unit (p. 43)
- [ ] Atoms and Molecules Unit (p. 69)
- [ ] Light Unit (p. 79)
- [ ] Sound Unit (p. 89)

Option 2 (By Season - Starting in the Fall)

- [ ] Solar System Unit (p. 43)
- [ ] Sound Unit (p. 89)
- [ ] Atoms and Molecules Unit (p. 69)
- [ ] Light Unit (p. 79)
- [ ] Major Biomes Unit (p. 31)
- [ ] Plants Unit (p. 13)
- [ ] Mendel Unit (p. 27)

If you have any questions or comments as you work through these materials, please don't hesitate to let us know by emailing support@elementalscience.com.

Table of Contents

Front Matter 1

Year A Overview 3
A Peek Inside a Science Chunks Unit 6
Introduction 7
Materials List 11

Plants Unit 13

Lesson 1: Leaves 14
Lesson 2: Flowers 16
Lesson 3: Fruits and Seeds 18
Lesson 4: Spores and Cones 20
Lesson 5: Stems 22
Lesson 6: Roots 24

Mendel Unit 27

Lesson 1: Gregor Mendel 28

Major Biomes Unit 31

Lesson 1: Polar Biome 32
Lesson 2: Forest Biome 34
Lesson 3: Grasslands Biome 36
Lesson 4: Desert Biome 38
Lesson 5: Aquatic Biome 40

Solar System Unit 43

Lesson 1: Our Solar System 44
Lesson 2: The Sun 46
Lesson 3: Mercury 48
Lesson 4: Venus 50
Lesson 5: The Earth and the Moon 52
Lesson 6: Mars 54
Lesson 7: Jupiter 56
Lesson 8: Saturn 58
Lesson 9: Uranus 60
Lesson 10: Neptune 62
Lesson 11: Dwarf Planets 64
Lesson 12: Asteroids, Comets, and Meteors 66

Atoms and Molecules Unit........................69

Lesson 1: Atoms 70

Lesson 2: Molecules 72

Lesson 3: Air 74

Lesson 4: Water 76

Light Unit........................79

Lesson 1: Light 80

Lesson 2: Color 82

Lesson 3: Light Behavior 84

Lesson 4: Lenses and Mirrors 86

Sound Unit........................89

Lesson 1: Sound 90

Lesson 2: Waves 92

Lesson 3: Wave Behavior 94

Lesson 4: Musical Instruments 96

Appendix........................99

Activity Sheet Template 100

Lab Report Template 101

Plant Growth Chart 102

Types of Roots Article 103

Biome Sheets 104

Tropical versus Temperate Template 109

Planet Templates for Projects 110

Paper-mâché Planet Directions 112

Phases of the Moon Template 113

Glossary........................115

Review Sheets........................121

Plants Review Sheet 122

Mendel Review Sheet 125

Major Biomes Review Sheet 127

Solar System Review Sheet 129

Atoms and Molecules Review Sheet 132

Light Review Sheet 133

Sound Review Sheet 135

A Peek Inside a Science Chunks Unit

1. Lesson Topic

Focus on one main idea throughout the week. You will learn about these ideas by reading from visually appealing encyclopedias, recording what the students learned, and doing coordinating hands-on science activities.

2. Information Assignments

Find two reading options—one for younger students, one for older students, plus optional library books.

3. Notebooking Assignments

Record what your students have learned with either a lapbook or a notebook. The directions for these options are included for your convenience in this section along with the vocabulary the lesson will cover.

4. Hands-on Science Assignments

Get the directions for coordinating hands-on science activities that relate to the week's topic.

5. Lesson To-Do Lists

See what is essential for you to do each week and what is optional. You can check these off as you work through the lesson so that you will know when you are ready to move on to the next one.

6. Lapbook Templates

Get all the information you need to create a lapbook on the subject. *(Printed bundle sold separately.)*

7. Notebook Templates

Have all the sheets you need to create a notebook on the subject, including a glossary for the vocabulary terms. *(Printed bundle sold separately.)*

In the appendix you will find a blank activity sheet, a blank lab report sheet, and any necessary templates.

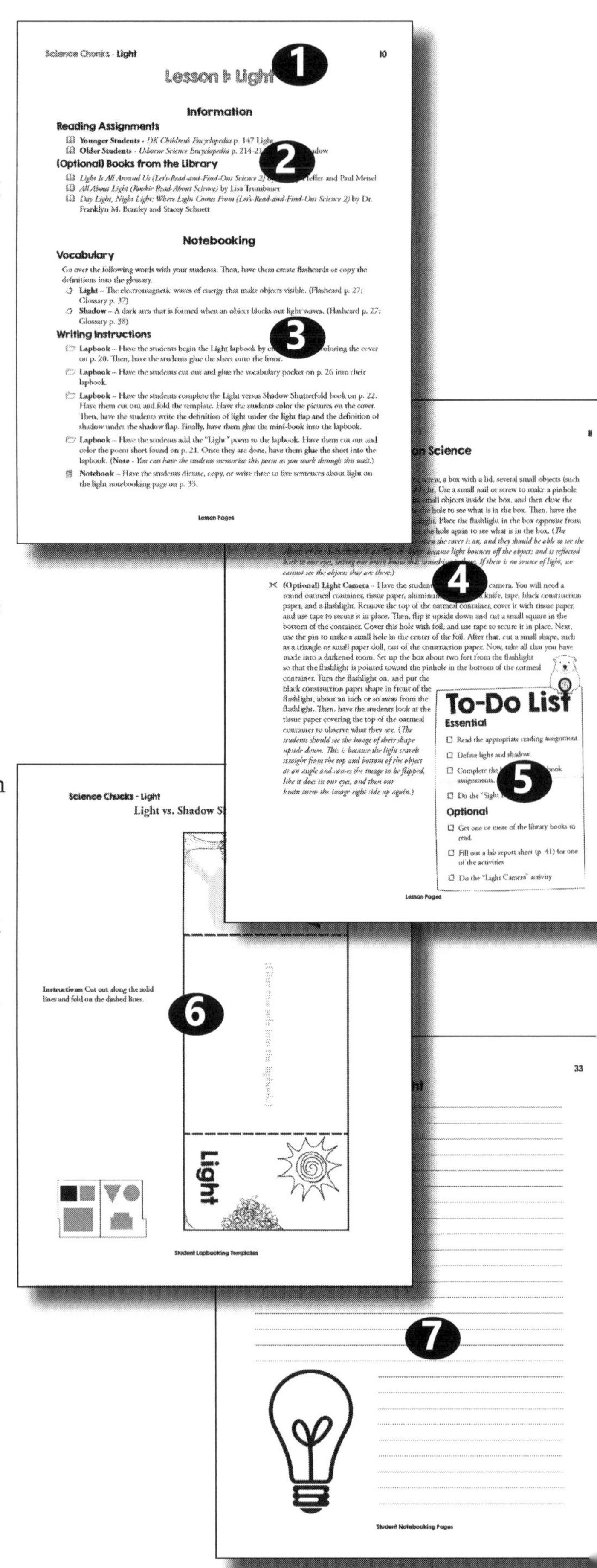

Introduction

Science Chunks - Year A is a unique and versatile unit study that leads you through a survey of plants, major biomes, solar system, atoms, light, and sound, plus a short unit on Gregor Mendel. It is designed to be a gentle approach to homeschool science based on the Unit Study method suggested in *Success in Science: A Manual for Excellence in Science Education* by Bradley and Paige Hudson. This study can be used as a stand-alone unit for elementary science.

What Is Included in Each Unit

Science Chunks - Year A includes seven units, each with the three keys to teaching science. For a lesson you will be doing the following:

✓ Listening to (or reading) **scientific information** from visually appealing encyclopedias

✓ Dictating (or writing down) what the students have learned and seen using **lapbooking or notebooking**

✓ Watching (and doing) **hands-on science** through a variety of science activities

Here is how this works.

Section I - Information

The elementary student is an empty bucket waiting to be filled with information, and science-oriented books are a wonderful way to do that. These books can include age-appropriate children's science encyclopedias, living books for science, and/or children's nonfiction science books.

In this program, the reading assignments and additional books scheduled in the lesson fulfill this component. The reading assignments are broken for you into two levels: younger students (1st to 3rd grade) and older students (4th to 6th grade).

Our idea is that you will read these selections with your students, pausing to ask questions or discussing the information once you are done reading.

Section 2 - Notebooking

The purpose of the notebooking component for elementary science education is to verify that the students have placed at least one piece of information into their knowledge bucket. You can use notebooking sheets, lapbooks, and/or vocabulary words to fulfill this requirement.

In this program, we have created two writing options, a lapbook and a notebook, for you to use with your students. In *Science Chunks Year A Student Lapbook Templates*, you will find all of the templates and pictures you will need to complete seven different lapbooks for the year. In *Science Chunks Year A Student Notebook*, you will find all the pages you need to create a simple

notebook documenting what the students learned this year, including notebooking sheets and a glossary for each unit.

Section 3 - Hands-on Science

Scientific demonstrations and observations are meant to spark students' enthusiasm for learning science, to work on their observation skills, and to demonstration the principles of science for them. This component of elementary science education can contain scientific demonstrations, hands-on projects, and/or nature studies.

In this program, the coordinating activities at the end of each lesson fulfill this section of elementary science instruction. If you would like to record what you have done, you can use one of the templates in the appendix pp. 100-101.

What You Need in Addition to This Guide

Books Scheduled

The following books are what we used to plan the reading assignments for this year:

- **Younger students (1st to 3rd grade)** - *Basher Biology, DK Children's Encyclopedia, Basher Chemistry*
- **Older students (4th to 6th grade)** - *Usborne Science Encyclopedia, Kingfisher Science Encyclopedia, DK Smithsonian Super Earth Encyclopedia*
- **All students** - *Gregor Mendel: The Friar Who Grew Peas by Cheryl Bardoe*

However, you could certainly use the encyclopedias you already have on hand or books from the library.

Unlike the single Science Chunks units, this printed Year A guide does not inlcude the student materials. You will need to purchase one (or both) of the following:

- *Science Chunks Year A Lapbook Templates (LT)*
- *Science Chunks Year A Student Notebook (NB)*

In addtion to the books and the student materials, you will need also simple craft supplies and other science materials—see a complete list of essential items on pp. 11-12.

How This Unit Works

We have included a to-do list with each lesson to give you an idea of what is essential and what is optional. There are several ways you can schedule this unit. Here is a quick look at a few of the options.

Possible Schedules for Your Week

- **One Day** – You can set aside about an hour to an hour and a half each week to complete all the essential tasks in one day.
- **Two Days** – You can set aside about 30 to 40 minutes twice a week to complete all the

essential tasks, plus a few more, in two days. On the first day, you can complete the reading assignments and either the lapbook or notebook assignments. On the second day, you can complete the coordinating activity and the vocabulary assignments as well as read any library books.

- **Three Days** – You can set aside about 30 minutes three times a week to complete all the essential tasks, plus a few more, in three days. On the first day, you can complete the reading assignments and either the lapbook or notebook assignments. On the second day, you can complete the coordinating activity and write a lab report using one of the templates. On the third day, you can do the vocabulary assignments as well as read any library books.
- **Four Days** – You can set aside about 20 to 30 minutes four times a week to complete all the essential tasks, plus a few more, in four days. On the first day, you can complete the reading assignments and either the lapbook or notebook assignments. On the second day, you can complete the coordinating activity and write a lab report. On the third day, you can do the vocabulary assignments as well as read any library books. On the fourth day, you can do the optional coordinating activity as well as read any library books.

Generally, we recommend completing one lesson per week, but please feel free to organize the units in the order that fits your goals.

Final Thoughts

Quick Links

View all the links mentioned in this guide in one place and get a digital copy of the appendix templates, glossary, and review sheets by visiting the following page:

https://elementalscience.com/blogs/resources/sc-a

Read Further

If you would like to read more about the philosophy behind the Science Chunks series, check out *Success in Science: A Manual for Excellence in Science Education* and the following articles from our website.

- **The Three Keys to Teaching Science** – This article shares the three keys to teaching science, including a free session that walks you through what each key can look like.

  https://elementalscience.com/blogs/news/3-keys
- **The Basics of Notebooking** – This article details the basic components of notebooking along with how a few suggestions on what notebooking can look like.

  https://elementalscience.com/blogs/news/what-is-notebooking
- **Lapbooking versus Notebooking** – This article takes a look at the differences between lapbooking and notebooking.

  https://elementalscience.com/blogs/news/lapbook-or-notebook

- **Scientific Demonstrations versus Experiments** – This article explains the difference between scientific demonstrations and experiments along with when and how to employ these methods.
    - https://elementalscience.com/blogs/news/89905795-scientific-demonstrations-or-experiments

Last Words

As the author and publisher of this curriculum, I encourage you to contact me with any questions or problems that you might have concerning *Science Chunks - Year A* by emailing us at support@elementalscience.com. I, or a member of our team, will be more than happy to answer them as soon as we can. I hope that you will enjoy creating memories this year!

~ Paige Hudson

Materials List

Plants Unit Materials

You will need the following materials to complete the essential coordinating activities:

- **Lesson 1:** Several different kinds of leaves (*try to include pine needles in the collection*), paper, and a crayon
- **Lesson 2:** A lily or other flower with clearly visible parts
- **Lesson 3:** Several types of fruit
- **Lesson 4:** A pine cone that is tightly closed and an oven
- **Lesson 5:** Jell-O, green jelly beans, grapes, a banana slice, a small ziploc bag, and a small square plastic container
- **Lesson 6:** Paper towel, bean seed, and plastic bag

Mendel Unit Materials

You will need the following materials to complete the essential coordinating activities:

- **Lesson 1:** LEGO® bricks in several sizes and colors

Major Biomes Unit Materials

You will need the following materials to complete the essential coordinating activities:

- **Lesson 1:** Epsom salts, hot water, food coloring, and paper
- **Lesson 2:** A 2-liter soda bottle with a top, gravel, potting soil, several small plants, scissors, tape, and water
- **Lesson 3:** Plastic or air-dry plants, real dirt, rocks, toy animal figures, and a shoebox
- **Lesson 4:** A shallow dish, sand, a few rocks, and several store-bought cacti or succulent plants
- **Lesson 5:** *No supplies needed*

Solar System Unit Materials

You will need the following materials to complete the essential coordinating activities:

- **Lesson 1:** Paper, string, and clothes hanger
- **Lesson 2:** Photo sensitive paper
- **Lesson 3:** Paint, picture of Mercury, and styrofoam ball or paper-mâché materials (balloon, newspaper, 1 cup of flour, ½ cup of water, and 2 tbsp of salt)
- **Lesson 4:** Paint, picture of Venus, and styrofoam ball or paper-mâché materials (balloon, newspaper, 1 cup of flour, ½ cup of water, and 2 tbsp of salt)
- **Lesson 5:** Eight sandwich-style cookies
- **Lesson 6:** Paint, picture of Mars, and styrofoam ball or paper-mâché materials (balloon, newspaper, 1 cup of flour, ½ cup of water, and 2 tbsp of salt)
- **Lesson 7:** Paint, picture of Jupiter, and styrofoam ball or paper-mâché materials (balloon, newspaper, 1 cup of flour, ½ cup of water, and 2 tbsp of salt)

- **Lesson 8:** Paint, picture of Saturn, and styrofoam ball or paper-mâché materials (balloon, newspaper, 1 cup of flour, ½ cup of water, and 2 tbsp of salt)
- **Lesson 9:** Paint, picture of Uranus, and styrofoam ball or paper-mâché materials (balloon, newspaper, 1 cup of flour, ½ cup of water, and 2 tbsp of salt)
- **Lesson 10:** Paint, picture of Neptune, and styrofoam ball or paper-mâché materials (balloon, newspaper, 1 cup of flour, ½ cup of water, and 2 tbsp of salt)
- **Lesson 11:** Paper and pencil or pen
- **Lesson 12:** Three feet of curling ribbon, a tennis ball, foil, and a straight pin

Atoms and Molecules Unit Materials

You will need the following materials to complete the essential coordinating activities:

- **Lesson 1:** 4 Pipe cleaners and round beads in three different colors (*at least 3 of each color*)
- **Lesson 2:** *No supplies needed.*
- **Lesson 3:** Balloon
- **Lesson 4:** Cup and several cubes of ice

Light Unit Materials

You will need the following materials to complete the essential coordinating activities:

- **Lesson 1:** A small nail or screw, a box with a lid, several small objects (such as a ball, a pencil, or a toy car), and a flashlight
- **Lesson 2:** A piece of paper, paint (red, yellow, and blue), pencil, and a paintbrush
- **Lesson 3:** An empty toilet paper roll, a thick mylar sheet, scissors, tape, card stock, a straw, and markers
- **Lesson 4:** A glass jar, water, pencil, and an index card

Sound Unit Materials

You will need the following materials to complete the essential coordinating activities:

- **Lesson 1:** *No supplies needed.*
- **Lesson 2:** A bowl of water and a small pebble
- **Lesson 3:** A string, popsicle sticks, ruler, pencil, and a hot glue gun
- **Lesson 4:** *No supplies needed.*

Abbreviations Key
LT = *Science Chunks Year A Student Lapbooking Templates*
NB = *Science Chunks Year A Student Notebook*

Plants Unit

Lesson 1: Leaves

Information

Reading Assignments

- **Younger Students** - *Basher Biology* p. 114 Leaves, p. 112 Chlorophyll
- **Older Students** - *Usborne Science Encyclopedia* p. 258-259 Leaves, p. 264 Plant Food

(Optional) Books from the Library

- *Why Do Leaves Change Color? (Let's-Read-and-Find... Science, Stage 2)* by Betsy Maestro
- *Leaves (Designs for Coloring)* by Ruth Heller
- *Leaf Jumpers* by Carole Gerber
- *Leaves* by David Ezra Stein
- *Photosynthesis: Changing Sunlight Into Food (Nature's Changes)* by Bobbie Kalman

Notebooking

Vocabulary

Go over the following word with your students. Then, have them create a flashcard or copy the definition into the glossary.

- **Leaf** – The part of the plant that makes the food for the plant. (Flashcard LT p. 17; Glossary NB p. 63)

Writing Instructions

- **Lapbook** – Have the students begin the Plants lapbook by cutting out and coloring the cover on LT p. 7. Then, have the students glue the sheet onto the front.
- **Lapbook (multi-week)** – Have the students cut out and color the cover page for the Parts of a Plant tab-book on LT p. 8 and the Leaves page on LT p. 9. Ask the students what they have learned about leaves this week and then add their narration to the leaves page of the Parts of Plants tab-book. Have them color the pictures on the leaves sheet. Save these two pages for when they assembles the tab-book in lesson 6 of this unit.
- **Lapbook** – Have the students cut out and color the "Parts of a Plant" poem on LT p. 12. Once they have finished, have them glue the poem into the lapbook.
- **Notebook** – Have the students dictate, copy, or write one to four sentences on what they have learned for leaves and photosynthesis on NB p. 6.

Hands-on Science

Coordinating Activities

- **Leaf Rubbings** – Have the students make a leaf rubbing booklet. Go on a nature walk and collect several different kinds of leaves – try to include pine needles in the collection. Once at home, use the samples to make a booklet of leaf rubbings. Begin this process by identifying the leaves you have collected. Then, place each leaf under a piece of paper and rub on the top of the same paper with a crayon until the shape of the leaf appears. Label the page with the type of leaf and set it aside. Once you have created a page for each of the leaves, bind the book together and create a cover.
- **(Optional) Plant Growth Project** – During this unit, you will record the growth of a bean plant. This week, begin this project by planting your seed. You will need dirt, a small pot, water, and a pinto bean seed. Fill the pot with dirt and gently press the bean seed just under the surface of the dirt. Water the pot well before placing it on a windowsill that receives direct sun light. Over the week, check your pot and water the plant when the soil is dry. At the end of your week, measure and record how much it has grown on the Plant Growth Record Chart on p. 102.

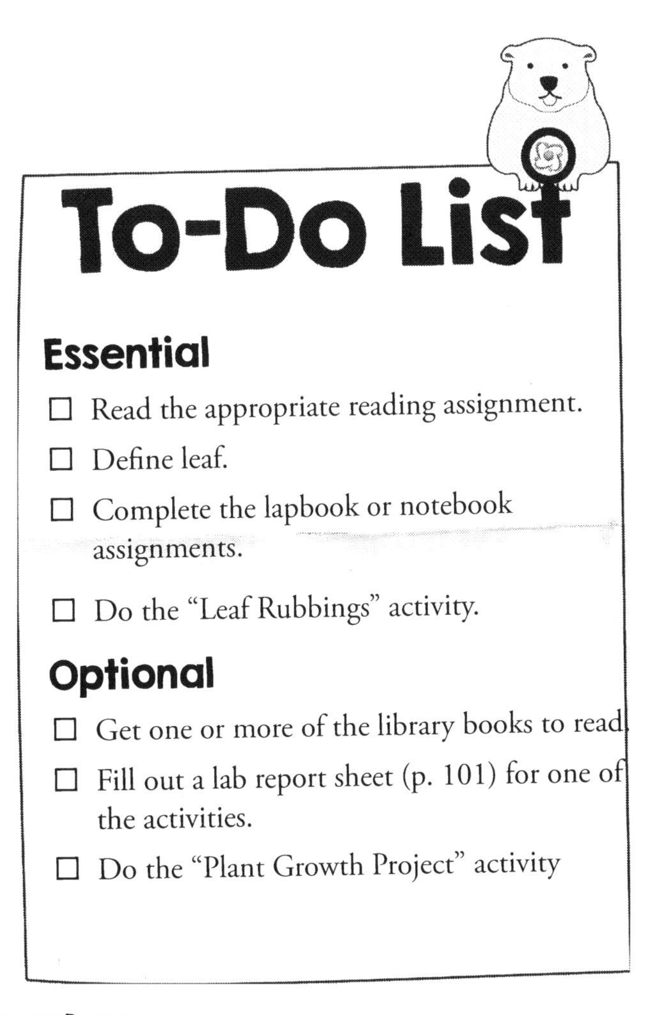

To-Do List

Essential

- ☐ Read the appropriate reading assignment.
- ☐ Define leaf.
- ☐ Complete the lapbook or notebook assignments.
- ☐ Do the "Leaf Rubbings" activity.

Optional

- ☐ Get one or more of the library books to read.
- ☐ Fill out a lab report sheet (p. 101) for one of the activities.
- ☐ Do the "Plant Growth Project" activity

Lesson 2: Flowers

Information

Reading Assignments

- **Younger Students** - *Basher Biology* p. 34 Flowering Plants, p. 118 Flower, p. 120 Pollen
- **Older Students** - *Usborne Science Encyclopedia* p. 270-271 Flowering Plants, part 1

(Optional) Books from the Library

- *The Reason for a Flower (World of Nature)* by Ruth Heller
- *A Weed Is a Flower* by Aliki
- *Flower (Life Cycle of A…)* by Molly Aloian

Notebooking

Vocabulary

Go over the following words with your students. Then, have them create a flashcard or copy the definition into the glossary.

- **Bud** – A swelling on a plant stem containing tiny flower part ready to burst into a bloom. (Flashcard LT p. 18; Glossary NB p. 59)
- **Flower** – The reproductive parts of a plant. (Flashcard LT p. 18; Glossary NB p. 61)

Writing Instructions

- **Lapbook** – Have the students cut out and color the pages of the Parts of a Flower mini tab-book on LT p. 13. Then, have them label the bud page with bud and stem, and the flower page with pistil, stigma, anther, stamen, and petals. Then, have them add what they learned about pollen to the pollen page. Finally, have the students staple the pages together and glue the mini tab-book into their lapbook. Here is an example of a labeled flower for your convenience.

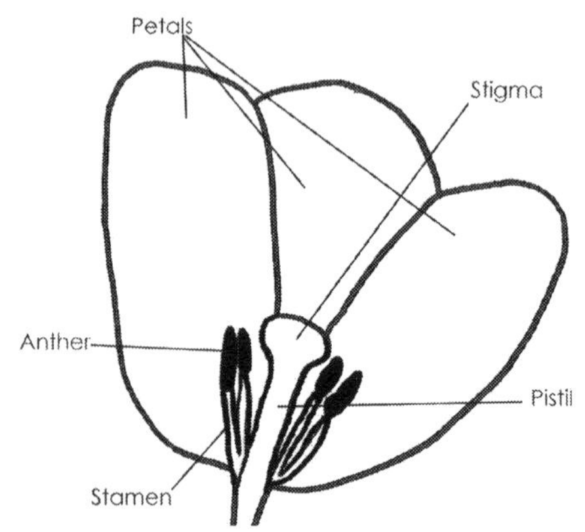

- **Lapbook (multi-week)** – Have the students cut out and color the flowers page of the Parts of Plants tab-book on LT p. 8. Ask the students what they have learned this week about flowers and then write their narration on the page. Save this page for when you assemble the tab-book in lesson 6.
- **Lapbook** – Have the students cut out and color the "Parts of a Flower" poem on LT p. 12. Once they have finished, have them glue the poem into the lapbook.

Notebook – Have the students dictate, copy, or write one to four sentences on what they have learned about flowering plants, flowers, and pollen on NB p. 7.

Hands-on Science

Coordinating Activities

Flower Dissection – Dissect a flower with the students. Purchase a lily or other flower with clearly visible parts. As you dissect the flower, be sure to point out the various parts to the students. For a more detailed explanation of this project, visit the following website:

- https://elementalscience.com/blogs/science-activities/94044099-how-to-dissect-a-flower

(Optional) Plant Growth Project – During this unit, you will record the growth of a bean plant. This week, water the plant as necessary. At the end of your week, measure and record how much it has grown on the Plant Growth Record Chart on p. 102.

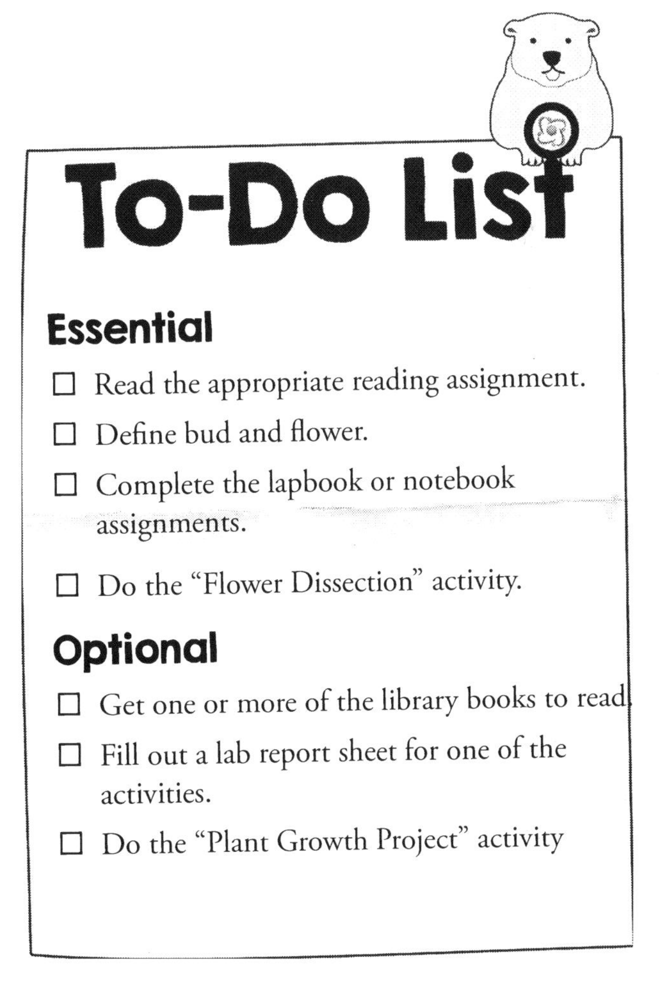

Lesson 3: Fruits and Seeds

Information

Reading Assignments

- **Younger Students** - *Basher Biology* p. 123 Fruit, p. 122 Seed
- **Older Students** - *Usborne Science Encyclopedia* p. 274 Seeds & Fruit

(Optional) Books from the Library

- *Seeds* by Ken Robbins
- *From Seed to Plant* by Gail Gibbons
- *A Fruit Is a Suitcase for Seeds* by Jean Richards
- *From Seed to Apple (How Living Things Grow)* by Anita Ganeri

Notebooking

Vocabulary

Go over the following word with your students. Then, have them create a flashcard or copy the definition into the glossary.

- **Seed** – The part of the plant that contains the beginnings of a new plant. (Flashcard LT p. 19; Glossary NB p. 68)

Writing Instructions

- **Lapbook** – Have the students cut out and color the Fruit mini-book on LT p. 14. Ask them what they have learned about fruit. Write their narration sentence on the inside of the book and have them glue the mini-book into their lapbook.
- **Lapbook**– Have the students cut out and color the seeds sheet on LT p. 14. Have them label the seed with radical and food store. Then, glue the back of the sheet to the lapbook. Here is an example of a labeled seed for your convenience:

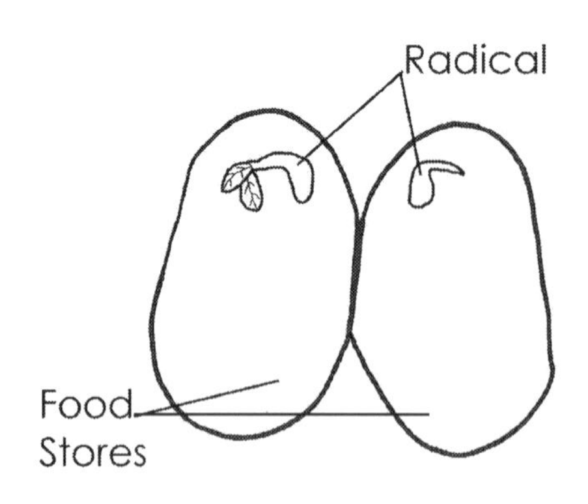

- **Notebook**– Have the students dictate, copy, or write one to four sentences on what they have learned about fruit and seeds on NB p. 8.

Hands-on Science

Coordinating Activities

- ✂ **Seeds in Fruit** – Plants have fruit to protect and distribute their seeds. Gather several types of fruit from your kitchen. Cut each one open and observe where they are found and what the seeds look like. Discuss the similarities and differences. Then, create a page that displays and identifies the different fruits and seeds you examined.
- ✂ **(Optional) Plant Growth Project** –During this unit, you will record the growth of a bean plant. This week, water the plant as necessary. At the end of your week, measure and record how much it has grown on the Plant Growth Record Chart on p. 102.

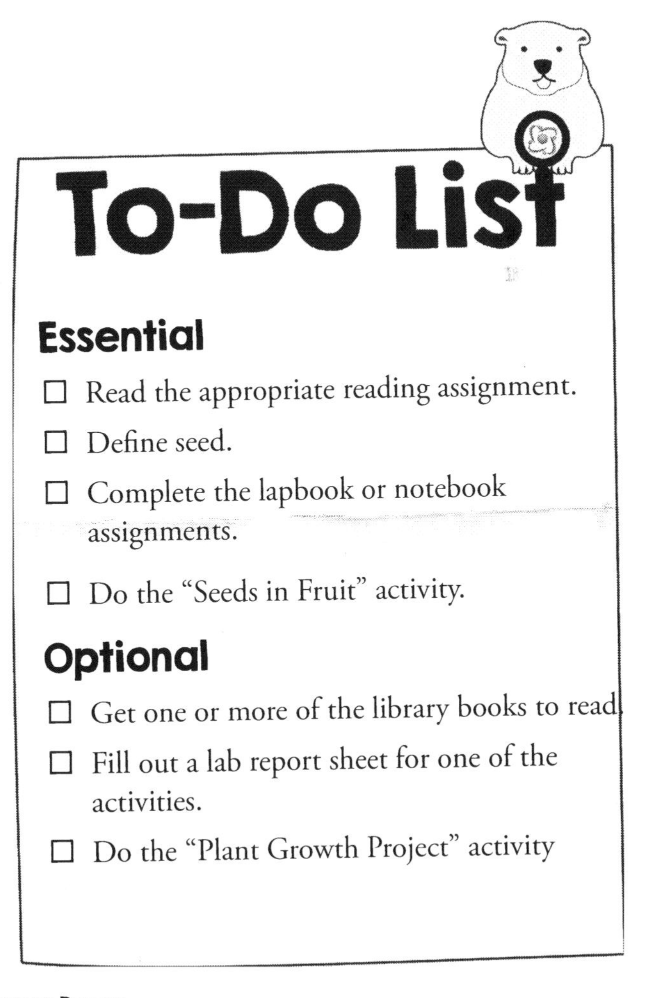

To-Do List

Essential

- ☐ Read the appropriate reading assignment.
- ☐ Define seed.
- ☐ Complete the lapbook or notebook assignments.
- ☐ Do the "Seeds in Fruit" activity.

Optional

- ☐ Get one or more of the library books to read.
- ☐ Fill out a lab report sheet for one of the activities.
- ☐ Do the "Plant Growth Project" activity

Lesson 4: Spores and Cones

Information

Reading Assignments

- **Younger Students** - *Basher Biology* p. 30 Seedless Plants, p. 32 Conifers
- **Older Students** - *Usborne Science Encyclopedia* p. 275 Cones, p. 282-283 Flowerless Plants

(Optional) Books from the Library

- *Plants That Never Ever Bloom (Explore!)* by Ruth Heller
- *Ferns (Rookie Read-About Science)* by Allan Fowler
- *From Pine cone to Pine Tree* by Ellen Weiss

Notebooking

Vocabulary

Go over the following word with your students. Then, have them create a flashcard or copy the definition into the glossary.

- **Cone** – A type of dry fruit produced by a conifer. (Flashcard LT p. 19; Glossary NB p. 59)

Writing Instructions

- **Lapbook** – Have the students cut out the pages of the Spores and Cones mini tab-book on LT p. 15. Ask them to tell you what they have learned about spores and cones. Write their narrations on each page of the tab-book. Have them color the pictures on each page before assembling the tab-book. Once the students have finished, glue the tab-book into their lapbook.

- **Notebook** – Have the students dictate, copy, or write one to four sentences on what they have learned for seedless plants and conifers on NT p. 9.

Hands-on Science

Coordinating Activities

- **Inside a Pine cone** – Have the students go outside to gather a pine cone that is tightly closed. Once inside, have them examine it. Ask the students these questions:

    - Is it hard or soft?

    - Is it sharp or dull?

  Then, gently heat the pine cone in a 300°F oven until the scales open up. (*This should take less than five minutes.*) Ask the students:

🗣 What do you see now?

After that, observe the pockets of space where the seeds normally rest. If there is a seed in one, take it out and examine it. For a more detailed explanation of this project, visit the following website:

🖱 http://elementalblogging.com/homeschool-science-corner-inside-cone/

✂ **(Optional) Plant Growth Project** –During this unit, you will record the growth of a bean plant. This week, water the plant as necessary. At the end of your week, measure and record how much it has grown on the Plant Growth Record Chart on p. 102.

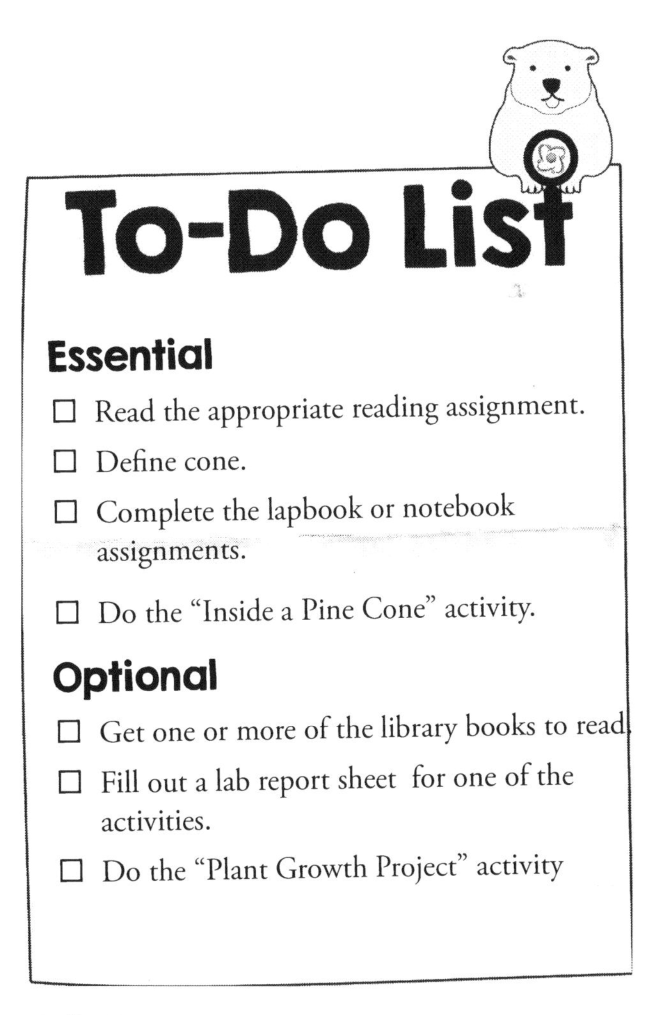

Lesson 5: Stems

Information

Reading Assignments

- **Younger Students** - *Basher Biology* p. 110 Plant Cell, p. 116 Stem
- **Older Students** - *Usborne Science Encyclopedia* p. 250-251 Plant Cells, p. 252 Stems

(Optional) Books from the Library

- *Stems (Plant Parts)* by Vijaya Bodach
- *Plant Plumbing: A Book About Roots and Stems* by Susan Blackaby
- *Stems (World of Plants)* by John Farndon

Notebooking

Vocabulary

Go over the following word with your students. Then, have them create a flashcard or copy the definition into the glossary.

- **Stem** – The part of a plant that holds it upright and supports its leaves and flowers. (Flashcard LT p. 20; Glossary NB p. 70)

Writing Instructions

- **Lapbook (multi-week)** – Have the students cut out and color the stem page of the Parts of Plants tab-book on LT p. 10. Ask the students what they have learned this week about stems and then write their narration on the page. Save this page for when you assemble the tab-book in lesson 6.
- **Lapbook** – Have the students complete the Plant Cell sheet on LT p. 11. Have them color the plant cell and label the cell wall, cell membrane, nucleus, and chloroplasts. You can use the diagram provided as a guide to help your students label the plant cell.
- **Notebook** – Have the students dictate, copy, or write one to four sentences on what they have learned for stems and plant cells on NB p. 10.

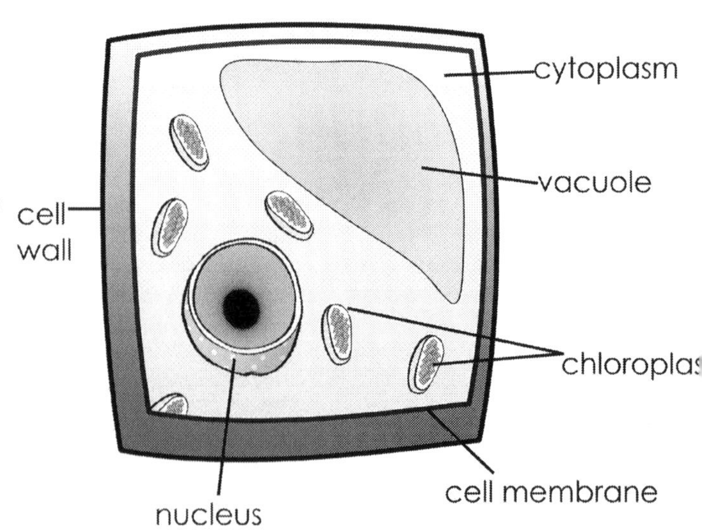

Hands-on Science

Coordinating Activities

- **Plant Cell** – Have the students make a model of a plant cell. They will need Jell-O, green jelly beans, grapes, a banana slice, a small Ziploc bag, and a small square plastic container. Begin by making the Jell-O according to the package directions and let it soft set. Have the students scoop a cup or two of lime Jell-O (the cytoplasm) into the Ziploc baggie (the cell membrane). Next, have them place the organelles into their cytoplasm. The banana slice will serve as the nucleus, the grapes as the vacuoles, and the jelly beans as the chloroplasts. Once the students have placed the different organelles, have them remove the air, seal the baggie, and place it in the square container which acts as the cell wall. Then, place their models back into the fridge for another hour or so to set. After the plant cell models have set, take them out and let the students observe their creations.
- **(Optional) Plant Growth Project** –During this unit, you will record the growth of a bean plant. This week, water the plant as necessary. At the end of your week, measure and record how much it has grown on the Plant Growth Record Chart on p. 102.

To-Do List

Essential

- ☐ Read the appropriate reading assignment.
- ☐ Define stem.
- ☐ Complete the lapbook or notebook assignments.
- ☐ Do the "Stems" activity.

Optional

- ☐ Get one or more of the library books to read.
- ☐ Fill out a lab report sheet for one of the activities.
- ☐ Do the "Plant Growth Project" activity

Lesson 6: Roots

Information

Reading Assignments

- **Younger Students** - *Basher Biology* p. 117 Roots
- **Older Students** - *Usborne Science Encyclopedia* p. 253 Roots, *Types of Roots article* on Appendix p. 199

(Optional) Books from the Library

- *What Do Roots Do?* by Kathleen V. Kudlinski
- *Roots (Plant Parts series)* by Vijaya Bodach

Notebooking

Vocabulary

Go over the following word with your students. Then, have them create a flashcard or copy the definition into the glossary.

- **Roots** – The part of the plant that anchors the plant firmly to the ground and absorbs water and nutrients. (Flashcard LT p. 20; Glossary NB p. 68)

Writing Instructions

- **Lapbook (multi-week)** – Have the students cut out and color the roots page of the Parts of Plants tab-book on LT p. 10. Ask the students what they have learned this week about roots and then write their narration on the page. Finally have them assemble the Parts of a Plant tab-book and glue the completed mini-book into their lapbook.
- **Lapbook** – Have the students complete the Types of Roots sheet on LT p. 16. Have them color the roots and label the top one with "fibrous root" and the bottom one with "taproot." You can also have the students add a sentence about each type under the picture.
- **Notebook** – Have the students dictate, copy, or write one to four sentences on what they have learned for roots and types of roots on NB p. 11.

Hands-on Science

Coordinating Activities

- **Growing Roots** – Have the students wet a paper towel with warm water and fold it in half. Take a bean seed from the kitchen and lay it on the wet paper towel. Place both inside a plastic bag, seal the baggie well. Set it in a warm, dark place for several days. Check the bag daily to see what happens. You should see the seed sprout and the roots begin to form.

✂ **(Optional) Plant Growth Project** – This is the final week for this project. Water the plant as necessary and then, At the end of your week, measure and record how much it has grown on the Plant Growth Record Chart on p. 102. After the students are finished, have them gently remove the plant from the pot and rinse the roots. Once it is clean, have them examine the root system of their bean plant up close.

Review Sheet

Plants Review Sheet Answers

1. Leaf
2. Photosynthesis
3. True
4. All
5. True
6. Protect and disperse
7. Dry fruits
8. True
9. Hold up flowers, Support the plant, Transport food and water
10. A. Cell wall, B. Chloroplasts, C. Nucleus
11. Fibrous root-grows out, Taproot-grows down
12. True

Essential

- ☐ Read the appropriate reading assignment.
- ☐ Define roots.
- ☐ Complete the lapbook or notebook assignments.
- ☐ Do the "Growing Roots" activity.

Optional

- ☐ Get one or more of the library books to read.
- ☐ Fill out a lab report sheet for one of the activities.
- ☐ Do the "Plant Growth Project" activity
- ☐ Complete the Plants Review Sheet. (p. 122-124)

Mendel Unit

Lesson 1: Gregor Mendel

Information

Reading Assignments

📖 **All Students** – *Gregor Mendel: The Friar Who Grew Peas* by Cheryl Bardoe (**Note**—You can read this book all in one sitting or break it up into two sessions by stopping at the page with three lilies that ends with, "If all went well, he would have a universal law that would apply to all living things.")

Discussion Questions

When you are done, you can ask your students the following questions:

- What led Gregor Mendel to study science? (Gregor Mendel wanted to study science because of what he had seen. He knew that growing two kinds of apple trees together produced better fruit and he wanted to know why.)
- How did Gregor Mendel get through school after his father injured his back? (Gregor Mendel worked as a tutor to pay for school and then later became a friar.)
- What are universal laws? (Universal laws explain why some things will always act in the same way, even in different settings.)
- What did Gregor Mendel believe about traits? (Gregor Mendel believed that plants and animals passed traits down from parents to their children.)
- What was Gregor Mendel's plan to test his theory? (Gregor Mendel's plan was to breed thousands of offspring from just a few parents, count how often the traits he was studying appeared, and see if he could use math to find a pattern.)
- What was Gregor's plan for his experiment? (He chose seven noticeable pairs of contrasting traits to track. He planned to breed each pair and record the offspring. Then, he would repeat the experiment until he had figured out the answer to his question.)
- How did Gregor Mendel control the breeding process? (Gregor Mendel controlled the breeding of his pea plants by fertilizing each flower by hand. Then, he covered each with a sack so that bees or butterflies could not contaminate his experiment.)
- What happened after the first breeding? (Gregor Mendel found that in each of the seven pairs of traits he choose, all of the children of the hybrid pairing looked like only one of the parents.) The second? (In the second trial, Mendel found that the missing traits reappeared.) The rest? (In subsequent trials, Mendel discovered that there was a pattern to how and when the traits were displayed.)
- What did Gregor Mendel discover? (Gregor Mendel discovered that some traits are dominant and some are recessive. He also found that the traits he studied acted independently.)
- What happened when he presented his findings? (When Gregor Mendel presented his

findings, no one really paid attention because they did not understand that his findings held value beyond peas.)

- **(Older Students)** Why do you think Gregor Mendel was willing to go hungry in order to continue his education? Would you be willing to do the same? (There is no right or wrong answer here; the point is to get your student thinking. If they are able to give good support to their argument, you have accomplished this goal.)
- **(Older Students)** Why do you think that it took decades for scientists to appreciate Mendel's work? (There is no right or wrong answer here; the point is to get your student thinking. If they are able to give good support to their argument, you have accomplished this goal.)

(Optional) Books from the Library

- *Gregor Mendel: Genetics Pioneer (Great Life Stories)* by Della Yannuzzi
- *Gregor Mendel: And the Roots of Genetics (Oxford Portraits in Science)* by Edward Edelson

Notebooking

Vocabulary

Go over the following words with your students. Then, have them create flashcards or copy the definitions into the glossary.

- **Dominant trait** – A characteristic that overrules the lesser seen recessive trait. (Flashcard LT p. 27; Glossary NB p. 60)
- **Genetic trait** – The characteristics, such as the color of a seed, that are inherited from a parent. (Flashcard LT p. 28; Glossary NB p. 62)
- **Hybrid** – A plant that is produced by crossbreeding. (Flashcard LT p. 28; Glossary NB p. 63)
- **Recessive trait** – The characteristic overruled by the dominant trait. (Flashcard LT p. 29; Glossary NB p. 67)

Writing Instructions

- **Lapbook** – Have the students complete the Gregor Mendel tab-book on LT p. 24-25. Have them cut out the mini-book pages and color the cover. Next, have the students tell you what they have learned about Gregor Mendel and his work on the respective tabs. Then, have them staple the pages of the booklet together and glue it into the lapbook.
- **Lapbook** – Have the students complete the Mendel timeline sheet on LT p. 26. Have them cut out the timeline sheet and color the pictures. Next, have the students add one to three of the significant dates (see below) on the timeline.
- **Lapbook** – Have the students cut out and glue the vocabulary pocket on LT p. 27 into their lapbook.
- **Notebook** – Have the students add the following dates to the Mendel timeline notebooking page on NB p. 16.
  - Gregor Mendel was born in 1822 in the Czech Republic.
  - In 1838, Gregor Mendel's father broke his back and left his son to fend for himself.

- In 1865, Gregor Mendel presented his findings to the Brno Natural History Society.
- Gregor Mendel died at the age of 63 in 1884 after a long illness.
- In 1900, three different scientists stumbled upon Mendel's findings and finally recognized the genius of his work.

Notebook – Have the students dictate, copy, or write information about Gregor Mendel and his work on the Mendel notebook pages on NB p. 14-15.

Hands-on Science

Coordinating Activity

LEGO® Punnett Squares – Have the students use a full LEGO® brick for the dominant trait and a half-brick for the recessive trait. Then, have them create several different Punnett Squares using the different options for parents (dominant - 2 full bricks; hybrid - one full brick and one half-brick; and recessive - 2 half-bricks). You can learn more about Punnett Squares, see an example of this project, and get a free worksheet at the following:

https://elementalscience.com/blogs/science-activities/punnett-square

(Optional) Plant Observation – Gregor Mendel observed his plants quite a bit. Through his regular observations, he learned to distinguish the different traits that he ultimately used in his experiment. In this activity the students will examine plants in their own environment to look for similarities and differences. You will need access to several different types of plants. Ideally, you will have up to 8 plants to observe. Begin by taking a walk outside, heading to your garden, or simply collecting the plants you observe. Have the students examine the size, the color, the texture, the leaf shape, any flowers or seeds, and the root structure if possible. If possible, have the students use the internet or a field guide to identify the plants they see.

To-Do List

Essential

- ☐ Read the appropriate reading assignment.
- ☐ Define dominant trait, genetic trait, hybrid, and recessive trait.
- ☐ Do the "Punnett Square" activity.
- ☐ Complete the lapbook or notebook assignments.

Optional

- ☐ Get one or more of the library books to read.
- ☐ Do the "Plant Observation" activity.
- ☐ Complete the Mendel Review Sheet on pp. 125-126.

Review Sheet

Mendel Review Sheet Answers

1. C, D, A, B
2. True
3. False (Gregor Mendel believed that plants and animals did pass traits down from parents to their children.)
4. False (Gregor Mendel did control the breeding of his pea plants.)
5. True
6. Answers will vary.

Major Biomes Unit

Lesson 1: Polar Biome

Information

Reading Assignments

- **Younger Students** – *DK Children's Encyclopedia* p. 126 Habitats, p. 197 Polar Habitats
- **Older Students** – *DK Smithsonian Super Earth Encyclopedia* pp. 200-201 Tundra, plus the following article on the difference between biome, ecosystem, and habitat:
  - https://elementalscience.com/blogs/science-activities/biome-ecosystem-habitat

(Optional) Books from the Library

- *Many Biomes, One Earth* by Sneed B. Collard III and James M. Needham
- *What Is a Biome? (Science of Living Things)* by Bobbie Kalman
- *The Arctic Habitat (Introducing Habitats)* by Molly Aloian and Bobbie Kalman
- *Arctic Tundra (Habitats)* by Michael H. Forman

Notebooking

Vocabulary

Go over the following words with your students. Then, have them create a flashcard or copy the definition into the glossary.

- **Biome** – A very large community of living things, both plants and animals. (Flashcard LT p. 39; Glossary NB p. 59)
- **Polar biome** – A biome with little vegetation and very cold temperatures. (Flashcard LT p. 40; Glossary NB p. 66

Writing Instructions

- **Lapbook** – Have the students begin the Major Biomes lapbook by cutting out and coloring the cover on LT p. 32. Then, have the students glue the sheet onto the front.
- **Lapbook** – Have the students complete the Biomes mini-book on LT p. 33. Have them cut out the template and color it. Next, have the students tell you what they have learned about biomes and write it on the inside of the booklet. Then,have them fold the book in half and glue the booklet into the lapbook.
- **Lapbook** – Have the students cut out the pages and color the cover of the Polar Biome flip-book on LT p. 34. Then, have the students color on the map where the polar biome habitat is typically found on the "Locations" page. After that, have the students tell you what they have learned about the polar biome and write it for them on the "Characteristics" page. Lastly, staple the pages together and glue the biome flip-book into the lapbook.
- **Notebook** – Have the students dictate, copy, or write two to six sentences on what they have learned for polar biome on NB p. 20.

Hands-on Science

Coordinating Activities

- **Ice Painting** – You will need Epsom salts, hot water, food coloring, and paper. Mix equal parts of the Epsom salts and hot water together until most of the Epsom salts have dissolved. Add a few drops of food coloring and mix well. Then, have the students use the mixture to paint a snowflake design or ice storm on the paper. As it dries, the ice crystals will form!
- **(Optional) Biomes of the World** – This week, you need coloring pencils and a picture of the polar biome or the template on p. 104. Have the students glue the photo of the polar biome on the poster, or have them color the picture on the template. You can also have them add a sentence or two about the biome to the poster.
- **(Optional) Memory Work** – There is no poem included in this lapbook as there is with other Science Chunks units. If you would still like to have your students work on memorizing a poem with this unit, you can use the following one:

Biomes
Deserts are dry and dusty places,
Hot all day, so water is scarce in these spaces.

The grassland is a prairie or pasture,
There are few trees and much grass for the horse and rancher.

The forest is full of different trees.
It has distinct layers that let plants grow with ease.

The arctic is a cold and icy land.
The ground is forever frozen, and the landscape is bland.

The aquatic zones are full of water.
Oceans and wetlands have fish and otter.

To-Do List

Essential

- ☐ Read the appropriate reading assignment.
- ☐ Define biome and polar biome.
- ☐ Complete the lapbook or notebook assignments.
- ☐ Do the "Ice Painting" activity.

Optional

- ☐ Get one or more of the library books to read.
- ☐ Fill out a lab report sheet (p. 101) for one of the activities.
- ☐ Do the "Biomes of the World" or the "Biomes Poem" activity.

Lesson 2: Forest Biome

Information

Reading Assignments

- **Younger Students** – *DK Children's Encyclopedia* p. 109 Forests, p. 204 Rainforests
- **Older Students** – *DK Smithsonian Super Earth Encyclopedia* pp. 190-191 Rainforest, pp.196-197 Temperate Forest, pp. 198-199 Boreal Forest

(Optional) Books from the Library

- *A Rainforest Habitat (Introducing Habitats)* by Molly Aloian
- *A Forest Habitat (Introducing Habitats)* by Bobbie Kalman
- *Northern Refuge: A Story of a Canadian Boreal Forest* by Audrey Fraggalosch

Notebooking

Vocabulary

Go over the following words with your students. Then, have them create a flashcard or copy the definition into the glossary.

- **Forest** – A biome characterized by an abundance of vegetation, especially trees. (Flashcard LT p. 40; Glossary NB p. 61)
- **Temperate zone** – Regions that do not experience extremes, so they have warm summers and cool winters. (Flashcard LT p. 41; Glossary NB p. 70)
- **Tropical zone** – Regions that are typically hot year-round. (Flashcard LT p. 41; Glossary NB p. 70)

Writing Instructions

- **Lapbook** – Have the students cut out the pages and color the cover of the Forest Biome flip-book on LT p. 35. Then, have the students color on the map where the forest biome habitat is typically found on the "Locations" page. After that, have the students tell you what they have learned about the forest biome and write it for them on the "Characteristics" page. Lastly, staple the pages together and glue the biome flip-book into the lapbook.
- **Notebook** – Have the students dictate, copy, or write two to six sentences on what they have learned for forest biome on NB p. 21.

Hands-on Science

Coordinating Activities

- **Rainforest in a Bottle** – Have the students recreate a rainforest in a bottle. You will need a 2-liter soda bottle with a top, gravel, potting soil, several small plants, scissors, tape, and water. Have an adult cut the soda bottle in half. Pour in a layer of gravel about 1 inch thick on the bottom of one half. Cover the gravel with a layer of potting soil several inches deep. Then, plant your plants and water them well. Tape the top half of the bottle back onto the bottom half, and place the "rainforest" in a bottle on a sunny window sill to observe what happens over the next several days. *(You should see that, after several hours, the bottle is coated with water droplets. Over several days, you will see that the soil remains moist and the bottle stays coated with water droplets. What is happening in the bottle is a small picture of the water cycle, which is repeated over and over in the rainforest.)*

✂ **(Optional) Tropical versus Temperate** – Have the students create a Venn diagram showing the differences and the similarities between tropical and temperate zones, such as in tropical and temperate forests. You can use the template in the appendix on p. 109. The following video will help your students as they complete this activity:

https://www.youtube.com/watch?v=3E2wfTB8YTg

✂ **(Optional) Biomes of the World** – This week, you need coloring pencils and a picture of the forest biome or the template on p. 105. Have the students glue the photo of the forest biome on the poster, or have them color the picture on the template. You can also have them add a sentence or two about the biome to the poster.

To-Do List

Essential

- ☐ Read the appropriate reading assignment.
- ☐ Define forest, temperate zone, and tropical zone.
- ☐ Complete the lapbook or notebook assignments.
- ☐ Do the "Rainforest in a Bottle" activity.

Optional

- ☐ Get one or more of the library books to read.
- ☐ Fill out a lab report sheet for one of the activities.
- ☐ Do the "Tropical versus Temperate" or the "Biomes of the World" activity

Lesson 3: Grasslands Biome

Information

Reading Assignments

- **Younger Students** – *DK Children's Encyclopedia* p. 125 Grasslands
- **Older Students** – *DK Smithsonian Super Earth Encyclopedia* pp. 192-193 Savanna

(Optional) Books from the Library

- *A Grassland Habitat (Introducing Habitats)* by Kelley Macaulay and Bobbie Kalman
- *Grasslands (About Habitats)* by Cathryn P. Sill
- *A Savanna Habitat (Introducing Habitats)* by Bobbie Kalman

Notebooking

Vocabulary

Go over the following word with your students. Then, have them create a flashcard or copy the definition into the glossary.

- **Grassland** – A biome with vast grassy fields. (Flashcard LT p. 42; Glossary NB p. 62)

Writing Instructions

- **Lapbook** – Have the students cut out the pages and color the cover of the Grasslands Biome flip-book on LT p. 36. Then, have the students color on the map where the grasslands biome habitat is typically found on the "Locations" page. After that, have the students tell you what they have learned about the grasslands biome and write it for them on the "Characteristics" page. Lastly, staple the pages together and glue the biome flip-book into the lapbook.
- **Notebook** – Have the students dictate, copy, or write two to six sentences on what they have learned for grasslands biome on NB p. 22.

Hands-on Science

Coordinating Activities

- **Habitat Diorama** – Have the students create a biome on the inside of a shoe box. They can choose from the three they have already studied—polar, forest, or grasslands—or choose the one of the two remaining biomes —desert or aquatic. Either way, the students can use plastic or air-dry plants along with real dirt and rocks. Alternatively, you could have them create the scene using construction paper. Be sure to have the students also place some of the animals typically found in the biome in their diorama as well.
- **(Optional) Biomes of the World** – This week, you need coloring pencils and a picture of

the grasslands biome or the template on p. 106. Have the students glue the photo of the grasslands biome on the poster, or have them color the picture on the template. You can also have them add a sentence or two about the biome to the poster.

To-Do List

Essential

- ☐ Read the appropriate reading assignment.
- ☐ Define grassland.
- ☐ Complete the lapbook or notebook assignments.
- ☐ Do the "Habitat diorama" activity.

Optional

- ☐ Get one or more of the library books to read.
- ☐ Fill out a lab report sheet for one of the activities.
- ☐ Do the "Biomes of the World" activity

Lesson 4: Desert Biome

Information

Reading Assignments

- **Younger Students** – *DK Children's Encyclopedia* p. 78 Desert
- **Older Students** – *DK Smithsonian Super Earth Encyclopedia* pp. 195-196 Desert

(Optional) Books from the Library

- *A Desert Habitat (Introducing Habitats)* by Kelley Macaulay and Bobbie Kalman
- *About Habitats: Deserts* by Cathryn P. Sill
- *Life in the Desert (Pebble Plus: Habitats Around the World)* by Alison Auch

Notebooking

Vocabulary

Go over the following words with your students. Then, have them create a flashcard or copy the definition into the glossary.

- **Desert** – The driest biome in the world. (Flashcard LT p. 42; Glossary NB p. 60)
- **Drought** – A long period without rain. (Flashcard LT p. 43; Glossary NB p. 60)

Writing Instructions

- **Lapbook** – Have the students cut out the pages and color the cover of the Desert Biome flip-book on LT p. 37. Then, have the students color on the map where the desert biome habitat is typically found on the "Locations" page. After that, have the students tell you what they have learned about the desert biome and write it for them on the "Characteristics" page. Lastly, staple the pages together and glue the biome flip-book into the lapbook.
- **Notebook** – Have the students dictate, copy, or write two to six sentences on what they have learned for desert biome on NB p. 23.

Hands-on Science

Coordinating Activities

- **Desert Garden** – Have the students make their own desert garden, using a shallow dish, sand, a few rocks, and several store-bought cacti or succulent plants.
- **(Optional) Biomes of the World** – This week, you need coloring pencils and a picture of the desert biome or the template on p. 107. Have the students glue the photo of the desert biome on the poster, or have them color the picture on the template. You can also have them add a

sentence or two about the biome to the poster.

To-Do List

Essential

- ☐ Read the appropriate reading assignment.
- ☐ Define desert and drought.
- ☐ Complete the lapbook or notebook assignments.
- ☐ Do the "Desert Garden" activity.

Optional

- ☐ Get one or more of the library books to read.
- ☐ Fill out a lab report sheet for one of the activities.
- ☐ Do the "Biomes of the World" activity

Lesson 5: Aquatic Biome

Information

Reading Assignments

- **Younger Students** – *DK Children's Encyclopedia* p. 187 Seas and Oceans
- **Older Students** – *DK Smithsonian Super Earth Encyclopedia* pp. 202-203 Marine Habitats

(Optional) Books from the Library

- *About Habitats: Wetlands* by Cathryn Sill and John Sill
- *A Wetland Habitat (Introducing Habitats)* by Molly Aloian and Bobbie Kalman
- *What Are Wetlands? (Science of Living Things)* by Bobbie Kalman and Amanda Bishop
- *Over in the Ocean: In a Coral Reef* by Marianne Berkes
- *Look Who Lives in the Ocean!: Splashing and Dashing, Nibbling and Quibbling, Blending and Fending* by Brooke Bessesen

Notebooking

Vocabulary

Go over the following word with your students. Then, have them create a flashcard or copy the definition into the glossary.

- **Aquatic biome** – A biome with an abundance of water, including oceans and wetlands. (Flashcard LT p. 43; Glossary NB p. 58)

Writing Instructions

- **Lapbook** – Have the students cut out the pages and color the cover of the Aquatic Biome flip-book on LT p. 38. Then, have the students label and color the world's major oceans (*Atlantic, Pacific, Arctic, Southern, and Indian Oceans*) on a map on the "World Oceans" page. After that, have the students tell you what they have learned about the aquatic biome and write it for them on the "Characteristics" page. Lastly, staple the pages together and glue the biome flip-book into the lapbook.
- **Notebook** – Have the students dictate, copy, or write two to six sentences on what they have learned for aquatic biome on NB p. 24.

Hands-on Science

Coordinating Activities

- **Wetlands Video** – Have the students watch the following video on the different types of wetlands:

  https://youtu.be/1WlmGyN9VXs

If possible, also take some time this week to go on a field trip with the students to a local river, lake, or ocean. Allow them to explore the environment, looking for animals, plants, and geographical features.

✂ **(Optional) Biomes of the World** – This week, you need coloring pencils and a picture of the aquatic biome or the template on p. 108. Have the students glue the photo of the aquatic biome on the poster, or have them color the picture on the template. You can also have them add a sentence or two about the biome to the poster.

Review Sheet

Major Biomes Review Sheet Answers

1. North
2. False *(The polar biome is the coldest biome on Earth.)*
3. Not all
4. Wettest
5. Hot, cool
6. Savannas
7. True
8. False *(Wetlands can have saltwater, fresh water, or brackish water.)*
9. C, A, E, B, D

To-Do List

Essential

☐ Read the appropriate reading assignment.
☐ Define aquatic biome.
☐ Complete the lapbook or notebook assignments.
☐ Do the "Wetlands" activity.

Optional

☐ Get one or more of the library books to read.
☐ Fill out a lab report sheet for one of the activities.
☐ Do the "Biomes of the World" activity
☐ Complete the Major Biomes Review Sheet. (p. 127-128)

Solar System Unit

Lesson 1: Our Solar System

Information

Reading Assignments

- **Younger Students** – *DK Children's Encyclopedia* p. 167 Milky Way, p. 233 Solar System
- **Older Students** – *Kingfisher Science Encyclopedia* pp. 390-391 Galaxies, pp. 398-399 The Solar System

(Optional) Books from the Library

- *The Milky Way (Exploring Space)* by Martha E. H. Rustad and Ilia I. Roussev
- *The Milky Way (Galaxy)* by Gregory L. Vogt
- *There's No Place Like Space: All About Our Solar System (Cat in the Hat's Learning Library)* by Tish Rabe and Aristides Ruiz
- *13 Planets: The Latest View of the Solar System (National Geographic Kids)* by David A. Aguilar
- *Scholastic Reader Level 2: Solar System* by Gregory Vogt
- *The Planets in Our Solar System (Let's-Read-and-Find... Science, Stage 2)* by Franklyn M. Branley and Kevin O'Malley

Notebooking

Vocabulary

Go over the following word with your students. Then, have them create a flashcard or copy the definition into the glossary.

- **Solar system** – A group of planets and other objects all in orbit around the Sun. (Flashcard LT p. 63; Glossary NB p. 69)

Writing Instructions

- **Lapbook** – Have the students begin the Solar System lapbook by cutting out and coloring the cover on LT p. 46. Then, have the students glue the sheet onto the front.
- **Lapbook** – Have the students cut out and color the "The Solar System" poem on LT p. 47. Once finished, have them glue the poem into the lapbook.
- **Lapbook** – Have the students cut out the template for the Solar System sheet on LT p. 47. Have them color the pictures and label each of the planets in our solar system. (*See the next page for a completed version.*) Have the students also fill in "Milky Way" for the name of our galaxy. Then, glue the sheet into the lapbook.
- **Notebook** – Have the students dictate, copy, or write one to four sentences on the Milky Way and our solar system on NB p. 26.

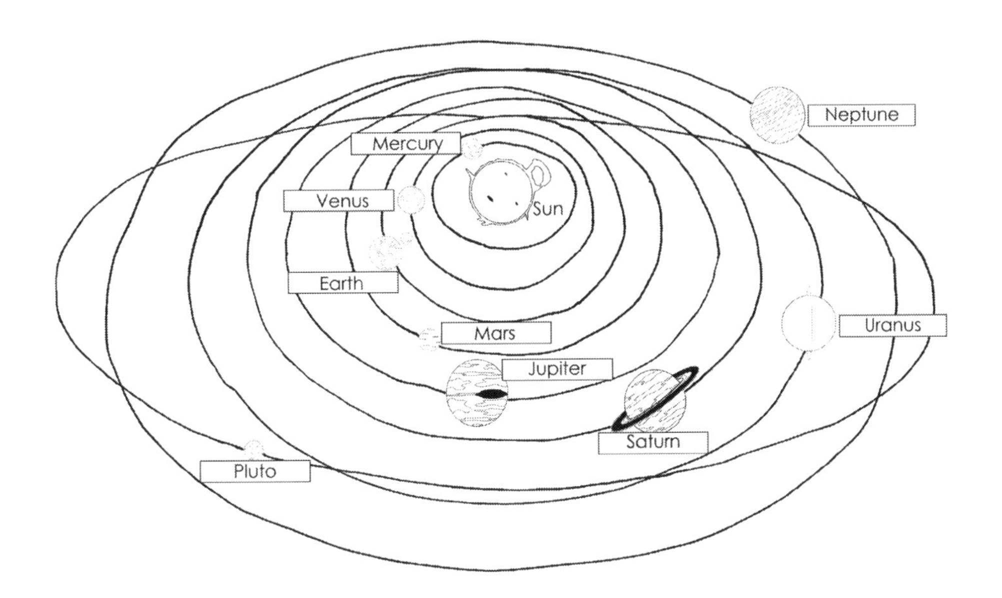

Hands-on Science

Coordinating Activities

- **Solar System Mobile** – Have the students make a hanger mobile of the solar system, using paper for your planets. Then, use string to attach the planets to a clothes hanger. You can use the planet templates in the appendix of this guide on pp. 110-111 for your project.
- **(Optional) Milky Way Art** – Have the students make their own Milky Way drawing using a white pastel or crayon on black construction paper. Then, have them use glue to trace the lines and sprinkle silver glitter over it.

To-Do List

Essential

- ☐ Read the appropriate reading assignment.
- ☐ Define solar system.
- ☐ Complete the lapbook or notebook assignments.
- ☐ Do the "Solar System Mobile" activity.

Optional

- ☐ Get one or more of the library books to read.
- ☐ Fill out a lab report sheet (p. 101) for one of the activities.
- ☐ Do the "Milky Way Art" activity

Lesson 2: The Sun

Information

Reading Assignments

- **Younger Students** – *DK Children's Encyclopedia* p. 247 Sun
- **Older Students** – *Kingfisher Science Encyclopedia* pp. 395-396 The Sun

(Optional) Books from the Library

- *The Sun: Our Nearest Star (Let's-Read-and-Find...)* by Franklyn M. Branley and Edward Miller
- *The Sun* by Seymour Simon
- *The Sun Is My Favorite Star* by Frank Asch
- *The Sun (Scholastic News Nonfiction Readers: Space Science)* by Melanie Chrismer

Notebooking

Vocabulary

Go over the following word with your students. Then, have them create a flashcard or copy the definition into the glossary.

- **Solar wind** – A stream of tiny particles that blow off the Sun and into space. (Flashcard LT p. 64, Glossary NB p. 69)

Writing Instructions

- **Lapbook** – Have the students cut out the template for the sun mini-book on LT p. 48 and fold along the dotted line. Have them tell you what they have learned about the sun and write it on the inside of the mini-book. Then, have the students color the picture and glue the mini-book into the lapbook.

- **Notebook** – Have the students dictate, copy, or write two to six total sentences on the Sun on NB p. 27. You can have older students label the sunspots and solar flare on the Sun.

Hands-on Science

Coordinating Activities

- **Sun Pictures** – Have the students use the sun to create a unique collage picture with photo sensitive paper, which is blue, but turns white when exposed to the sun. It can be purchased at your local craft store. If you cover a portion of it, that part will remain blue, thus creating a picture. Have the student lay out his design on the paper according to the directions that come with it. Then, lay the paper out in the sun and watch the creation develop.
- **(Optional) Solar System Wall Model** – Have the students create a model of the solar

system on a wall in your school room. This week, have them create a large paper version of the Sun. Begin by taping together three sheets of yellow construction paper. Then, have the students trace a portion of a semi-circle. Have the students add details like sunspots and solar flares before cutting out the Sun. Place the model on the far left of the wall on which plan to make your model.

To-Do List

Essential

- ☐ Read the appropriate reading assignment.
- ☐ Define solar wind.
- ☐ Complete the lapbook or notebook assignments.
- ☐ Do the "Sun Pictures" activity.

Optional

- ☐ Get one or more of the library books to read.
- ☐ Fill out a lab report sheet for one of the activities.
- ☐ Do the "Solar System Wall Model" activity

Lesson 3: Mercury

Information

Reading Assignments

- **Younger Students** – *DK Children's Encyclopedia* p. 161 Mercury
- **Older Students** – *Kingfisher Science Encyclopedia* p. 403 Mercury

(Optional) Books from the Library

- *Mercury (True Books: Space)* by Elaine Landau
- *Mercury (First Facts: Solar System)* by Adele Richardson
- *Mercury (Early Bird Astronomy)* by Gregory L. Vogt
- *Mercury (Mulberry books)* by Seymour Simon

Notebooking

Vocabulary

Go over the following word with your students. Then, have them create a flashcard or copy the definition into the glossary.

- **Planet** – A large ball of rock or gas that travels around a star. (Flashcard LT p. 64, Glossary NB p. 66)

Writing Instructions

- **Lapbook** – Have the students complete the Mercury tab-book on LT p. 49. Have them cut out the template and color the cover. Have them color in the location of the planet in our solar system and fill in the planet details. Then, have the students tell you what they have learned about Mercury and write it on the last page of the tab-book. Finally, have them glue the tab-book into the lapbook.
- **Lapbook** – Have the students cut out and color the "Planet" poem on LT p. 50. Once finished, have them glue the poem into the lapbook.
- **Notebook** – Have the students dictate, copy, or write two to six total sentences on Mercury on NB p. 28. You can have older students label the Caloris Basin and craters on Mercury.

Hands-on Science

Coordinating Activities

- **Mercury Model** – Have the students make a model of Mercury for a solar system model. You can choose to make the model out of Styrofoam balls or paper-mâché. Instructions for making for making paper-mâché planets are found in the appendix on p. 112.

✂ **(Optional) Solar System Wall Model** – This week, have the students add Mercury to their solar system wall model. See the following chart for the size of the planet and its distance from the Sun.

Planet	Distance from the Sun	Scale diameter of planet
Mercury	2 inches	3/4 inch

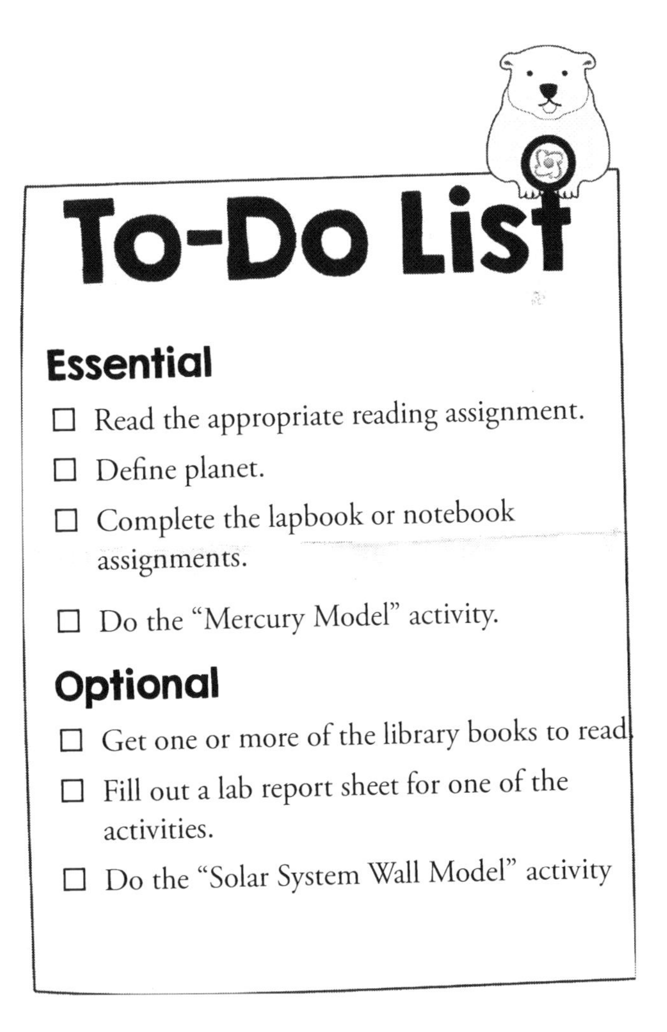

Lesson 4: Venus

Information

Reading Assignments

- **Younger Students** – *DK Children's Encyclopedia* p. 265 Venus
- **Older Students** – *Kingfisher Science Encyclopedia* p. 404 Venus

(Optional) Books from the Library

- *Venus (True Books: Space)* by Elaine Landau
- *Venus (First Facts: Solar System)* by Adele Richardson
- *Venus* by Seymour Simon

Notebooking

Vocabulary

Go over the following word with your students. Then, have them create a flashcard or copy the definition into the glossary.

- **Atmosphere** – A layer of gases that surrounds a planet. (Flashcard LT p. 65, Glossary NB p. 58)

Writing Instructions

- **Lapbook** – Have the students complete the Venus tab-book on LT p. 51. Have them cut out the template and color the cover. Have them color in the location of the planet in our solar system and fill in the planet details. Then, have the students tell you what they have learned about Venus and write it on the last page of the tab-book. Finally, have them glue the tab-book into the lapbook.
- **Notebook** – Have the students dictate, copy, or write two to six total sentences on Venus on NB p. 29. You can have older students label the layers of thick clouds on Venus.

Hands-on Science

Coordinating Activities

- **Venus Model** – Have the students make a model of Venus for a solar system model. You can choose to make the model out of Styrofoam balls or paper-mâché. Instructions for making a wall model and for making paper-mâché planets are found in the appendix on p. 112.
- **(Optional) Solar System Wall Model** – This week, have the students add Venus to their solar system wall model. See the following chart for the size of the planet and its distance from the Sun.

Planet	Distance from the Sun	Scale diameter of planet
Venus	3 inches	1 3/4 inches

To-Do List

Essential

- ☐ Read the appropriate reading assignment.
- ☐ Define atmosphere.
- ☐ Complete the lapbook or notebook assignments.
- ☐ Do the "Venus Model" activity.

Optional

- ☐ Get one or more of the library books to read.
- ☐ Fill out a lab report sheet for one of the activities.
- ☐ Do the "Solar System Wall Model" activity

Lesson 5: The Earth and the Moon

Information

Reading Assignments

- **Younger Students** – *DK Children's Encyclopedia* p. 83 Earth, p. 171 Moon
- **Older Students** – *Kingfisher Science Encyclopedia* pp. 400-401 Earth and the Moon

(Optional) Books from the Library

- *Inside The Earth (Magic School Bus)* by Joanna Cole and Bruce Degen
- *Earth (Space)* by Charlotte Guillain
- *Earth (Scholastic News Nonfiction Readers: Space Science)* by Christine Taylor-Butler
- *Faces of the Moon* by Bob Crelin and Leslie Evans
- *The Moon Book* by Gail Gibbons
- *The Moon Seems to Change (Let's-Read-and-Find... Science 2)* by Franklyn M. Branley and Barbara and Ed Emberley

Notebooking

Vocabulary

Go over the following word with your students. Then, have them create a flashcard or copy the definition into the glossary.

- **Moon** – A mini-planet in orbit around another planet. (Flashcard LT p. 65, Glossary NB p. 65)

Writing Instructions

- **Lapbook** – Have the students complete the Earth tab-book on LT p. 52. Have them cut out the template and color the cover. Have them color in the location of the planet in our solar system and fill in the planet details. Then, have the students tell you what they have learned about Earth and write it on the last page of the tab-book. Finally, have them glue the tab-book into the lapbook.
- **Lapbook** – Have the students complete the Moon mini-book on LT p. 53. Have them cut out the template and fold along the dotted line. Have them tell you what they have learned about the Moon and write it on the inside of the mini-book. Then, have them color the picture and glue the mini-book into the lapbook.
- **Notebook** – Have the students dictate, copy, or write two to six total sentences on Earth and its Moon on NB p. 30.

Hands-on Science

Coordinating Activities

- **Phases of the moon** – Have the students explore the phases of the Moon through cookies. You will need eight sandwich-style cookies and a sheet with the phases of the Moon on it (see p. 113). Have the student open up the cookies and remove a portion of the icing to match the different phases of the Moon on the sheet.
- **(Optional) Solar System Wall Model** – This week, have the students add Earth to their solar system wall model. Below is a chart they can use for the size of the planet and its distance from the Sun.

Planet	Distance from the Sun	Scale diameter of planet
Earth	4 inches	2 inches

The students can also add the Moon, which would be one-quarter of the size of the Earth, to their wall model.

To-Do List

Essential

- ☐ Read the appropriate reading assignment.
- ☐ Define moon.
- ☐ Complete the lapbook or notebook assignments.
- ☐ Do the "Phases of the Moon" activity.

Optional

- ☐ Get one or more of the library books to read
- ☐ Fill out a lab report sheet for one of the activities.
- ☐ Do the "Solar System Wall Model" activity

Lesson 6: Mars

Information

Reading Assignments

- **Younger Students** – *DK Children's Encyclopedia* p. 156 Mars
- **Older Students** – *Kingfisher Science Encyclopedia* p. 405 Mars

(Optional) Books from the Library

- *You Are the First Kid on Mars* by Patrick O'Brien
- *Mission to Mars (Let's-Read-and-Find... Science 2)* by Franklyn M. Branley and True Kelley
- *Mars: The Red Planet (Our Solar System)* by Lincoln James
- *Destination: Mars* by Seymour Simon
- *Mars (Scholastic News Nonfiction Readers: Space Science)* by Melanie Chrismer
- *Planet Mars (True Books)* by Ann O. Squire

Notebooking

Vocabulary

Go over the following word with your students. Then, have them create a flashcard or copy the definition into the glossary.

- **Orbit** – The path of an object in space. (Flashcard LT p. 66, Glossary NB p. 66)

Writing Instructions

- **Lapbook** – Have the students complete the Mars tab-book on LT p. 54. Have them cut out the template and color the cover. Have them color in the location of the planet in our solar system and fill in the planet details. Then, have the students tell you what they have learned about Mars and write it on the last page of the tab-book. Finally, have them glue the tab-book into the lapbook.
- **Lapbook** – Have the students add the Inner Planets label on LT p. 55. Have them cut out and color the inner planets. You can also have the students label each planet. Once they finish, have them glue it into their lapbook.
- **Notebook** – Have the students dictate, copy, or write two to six total sentences on Mars on NB p. 31. You can have older students label the polar regions on Mars.

Hands-on Science

Coordinating Activities

- **Mars Model** – Have the students make a model of Mars for a solar system model. You can choose to make the model out of Styrofoam balls or paper-mâché. Instructions for making a wall model and for making paper-mâché planets are found in the appendix on p. 112.

✂ **(Optional) Solar System Wall Model** – This week, have the students add Mars to their solar system wall model. See the following chart for the size of the planet and its distance from the Sun.

Planet	Distance from the Sun	Scale diameter of planet
Mars	6 inches	1 1/8 inches

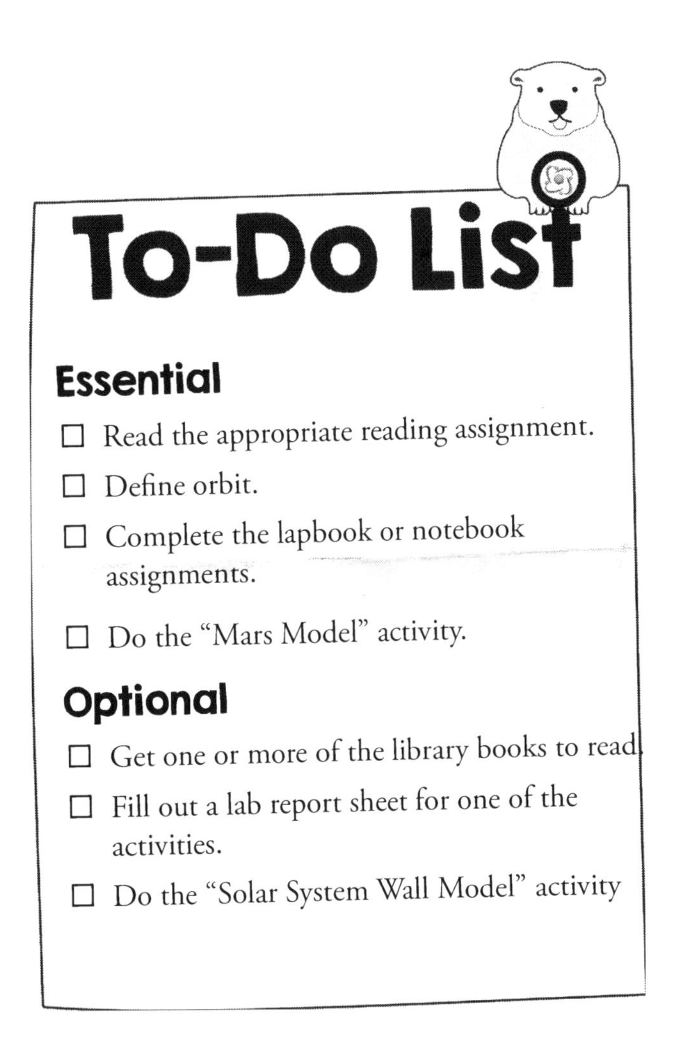

Lesson 7: Jupiter

Information

Reading Assignments

- **Younger Students** – *DK Children's Encyclopedia* p. 141 Jupiter
- **Older Students** – *Kingfisher Science Encyclopedia* p. 406 Jupiter

(Optional) Books from the Library

- *Destination: Jupiter* by Seymour Simon
- *The Largest Planet: Jupiter (Amazing Science: Planets)* by Nancy Loewen and Jeff Yesh
- *Jupiter: The Largest Planet (Our Solar System)* by Daisy Allyn
- *Planet Jupiter (True Books)* by Ann O. Squire
- *Jupiter (Scholastic News Nonfiction Readers: Space Science)* by Christine Taylor-Butler
- *Jupiter (Blastoff! Readers: Exploring Space)* by Derek Zobel

Notebooking

Vocabulary

Go over the following word with your students. Then, have them create a flashcard or copy the definition into the glossary.

- **Gravity** – The force that pulls objects, such as the Sun and Jupiter, towards other objects. (Flashcard LT p. 66, Glossary NB p. 62)

Writing Instructions

- **Lapbook** – Have the students complete the Jupiter tab-book on LT p. 56. Have them cut out the template and color the cover. Have them color in the location of the planet in our solar system and fill in the planet details. Then, have the students tell you what they have learned about Jupiter and write it on the last page of the tab-book. Finally, have them glue the tab-book into the lapbook.

- **Notebook** – Have the students dictate, copy, or write two to six total sentences on Jupiter on NB p. 32. You can have older students label the bands of storms and the red spot.

Hands-on Science

Coordinating Activities

- **Jupiter Model** – Have the students make a model of Jupiter for a solar system model. You can choose to make the model out of Styrofoam balls or paper-mâché. Instructions for making a wall model and for making paper-mâché planets are found in the appendix on p. 112.

✂ **(Optional) Solar System Wall Model** – This week, have the students add Jupiter to their solar system wall model. See the following chart for the size of the planet and its distance from the Sun.

Planet	Distance from the Sun	Scale diameter of planet
Jupiter	1 foot, 9 inches	22 inches

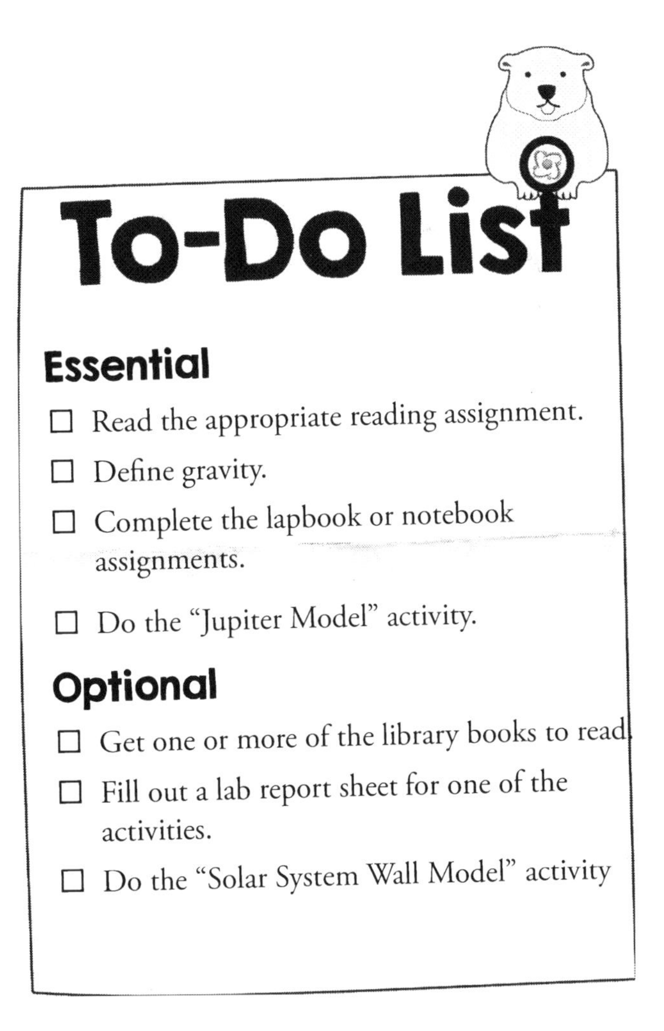

Lesson 8: Saturn

Information

Reading Assignments

- **Younger Students** – *DK Children's Encyclopedia* p. 216 Saturn
- **Older Students** – *Kingfisher Science Encyclopedia* p. 407 Saturn

(Optional) Books from the Library

- *Saturn* by Seymour Simon
- *Saturn (True Books: Space)* by Elaine Landau
- *Saturn: The Ringed Planet (Our Solar System)* by Daisy Allyn
- *Planet Saturn (True Books)* by Ann O. Squire
- *Saturn (Scholastic News Nonfiction Readers: Space Science)* by Christine Taylor-Butler
- *Jupiter and Saturn (Up in Space)* by Rosalind Mist

Notebooking

Vocabulary

- There is no vocabulary for this lesson.

Writing Instructions

- **Lapbook** – Have the students complete the Saturn tab-book on LT p. 57. Have them cut out the template and color the cover. Have them color in the location of the planet in our solar system and fill in the planet details. Then, have the students tell you what they have learned about Saturn and write it on the last page of the tab-book. Finally, have them glue the tab-book into the lapbook.
- **Notebook** – Have the students dictate, copy, or write two to six total sentences on Saturn on NB p. 33. You can have older students label the rings.

Hands-on Science

Coordinating Activities

- **Saturn Model** – Have the students make a model of Saturn for a solar system model. You can choose to make the model out of Styrofoam balls or paper-mâché. Instructions for making a wall model and for making paper-mâché planets are found in the appendix on p. 112.
- **(Optional) Solar System Wall Model** – This week, have the students add Saturn to their solar system wall model. See the following chart for the size of the planet and its distance

from the Sun.

Planet	Distance from the Sun	Scale diameter of planet
Saturn	3 feet, 2 inches	20 inches

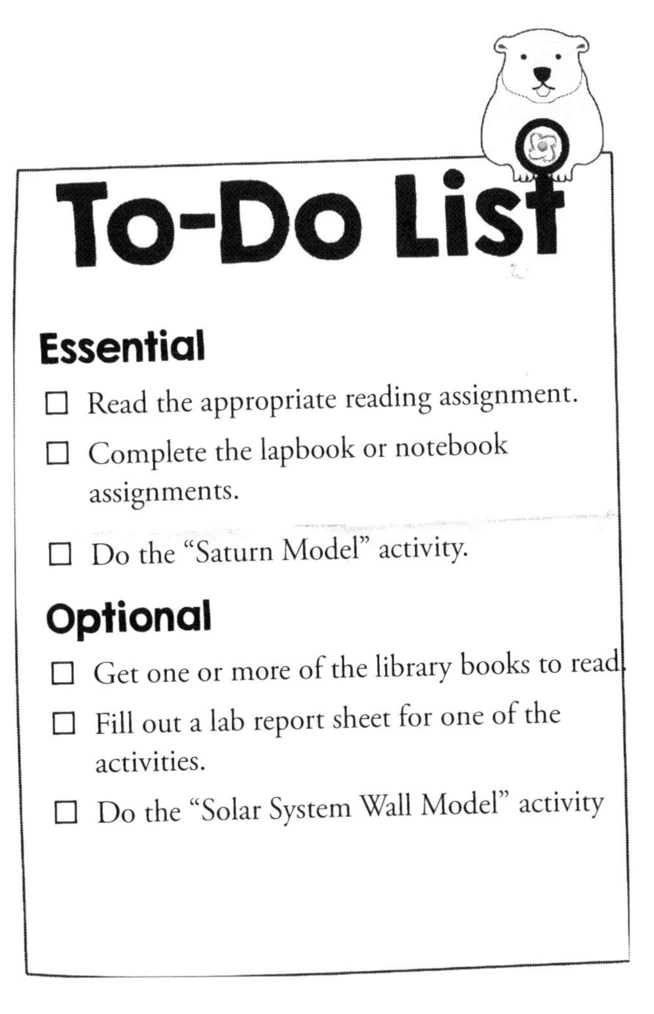

Lesson 9: Uranus

Information

Reading Assignments

- **Younger Students** – *DK Children's Encyclopedia* p. 264 Uranus
- **Older Students** – *Kingfisher Science Encyclopedia* p. 408 Uranus

(Optional) Books from the Library

- *Uranus (True Books)* by Christine Taylor-Butler
- *Uranus (True Books: Space)* by Elaine Landau
- *Uranus* by Seymour Simon
- *Uranus: The Ice Planet (Our Solar System)* by Greg Roza
- *The Sideways Planet: Uranus (Amazing Science: Planets)* by Nancy Loewen and Jeff Yesh

Notebooking

Vocabulary

- There is no vocabulary for this lesson.

Writing Instructions

- **Lapbook** – Have the students complete the Uranus tab-book on LT p. 58. Have them cut out the template and color the cover. Have them color in the location of the planet in our solar system and fill in the planet details. Then, have the students tell you what they have learned about Uranus and write it on the last page of the tab-book. Finally, have them glue the tab-book into the lapbook.
- **Notebook** – Have the students dictate, copy, or write two to six total sentences on Uranus on NB p. 34. You can have older students label the rings.

Hands-on Science

Coordinating Activities

- **Uranus Model** – Have the students make a model of Uranus for a solar system model. You can choose to make the model out of Styrofoam balls or paper-mâché. Instructions for making a wall model and for making paper-mâché planets are found in the appendix on p. 112.
- **(Optional) Solar System Wall Model** – This week, have the students add Uranus to their solar system wall model. See the following chart for the size of the planet and its distance

from the Sun.

Planet	Distance from the Sun	Scale diameter of planet
Uranus	6 feet, 5 inches	8 inches

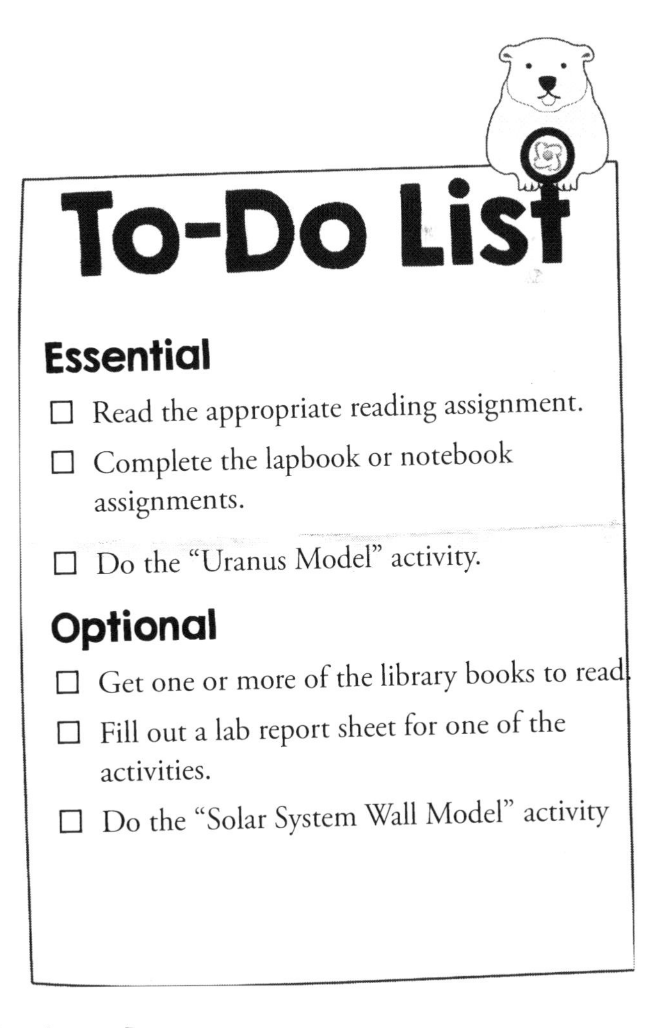

Lesson 10: Neptune

Information

Reading Assignments

- **Younger Students** – *DK Children's Encyclopedia* p. 183 Neptune
- **Older Students** – *Kingfisher Science Encyclopedia* p. 409 Neptune

(Optional) Books from the Library

- *Neptune* by Seymour Simon
- *Neptune (True Books: Space)* by Elaine Landau
- *Neptune: The Stormy Planet (Our Solar System)* by Greg Roza
- *Planet Neptune (True Books)* by Ann O. Squire
- *Neptune (Scholastic News Nonfiction Readers: Space Science)* by Melanie Chrismer
- *Farthest from the Sun: The Planet Neptune (Amazing Science: Planets)* by Nancy Loewen and Jeff Yesh

Notebooking

Vocabulary

- There is no vocabulary for this lesson.

Writing Instructions

- **Lapbook** – Have the students complete the Neptune tab-book on LT p. 59. Have them cut out the template and color the cover. Have them color in the location of the planet in our solar system and fill in the planet details. Then, have the students tell you what they have learned about Neptune and write it on the last page of the tab-book. Finally, have them glue the tab-book into the lapbook.
- **Lapbook** – Have the students add the Outer Planets label on LT p. 55. Have them cut out and color the outer planets. You can also have the students label each planet. Once they finish, have them glue it into their lapbook.
- **Notebook** – Have the students dictate, copy, or write two to six total sentences on Neptune on NB p. 35. You can have older students label the rings.

Hands-on Science

Coordinating Activities

- **Neptune Model** – Have the students make a model of Neptune for a solar system model. You can choose to make the model out of Styrofoam balls or paper-mâché. Instructions for making a wall model and for making paper-mâché planets are found in the appendix on p. 112.

✂ **(Optional) Solar System Wall Model** – This week, have the students add Neptune to their solar system wall model. See the following chart for the size of the planet and its distance from the Sun.

Planet	Distance from the Sun	Scale diameter of planet
Neptune	10 feet, 1 inch	7 1/2 inches

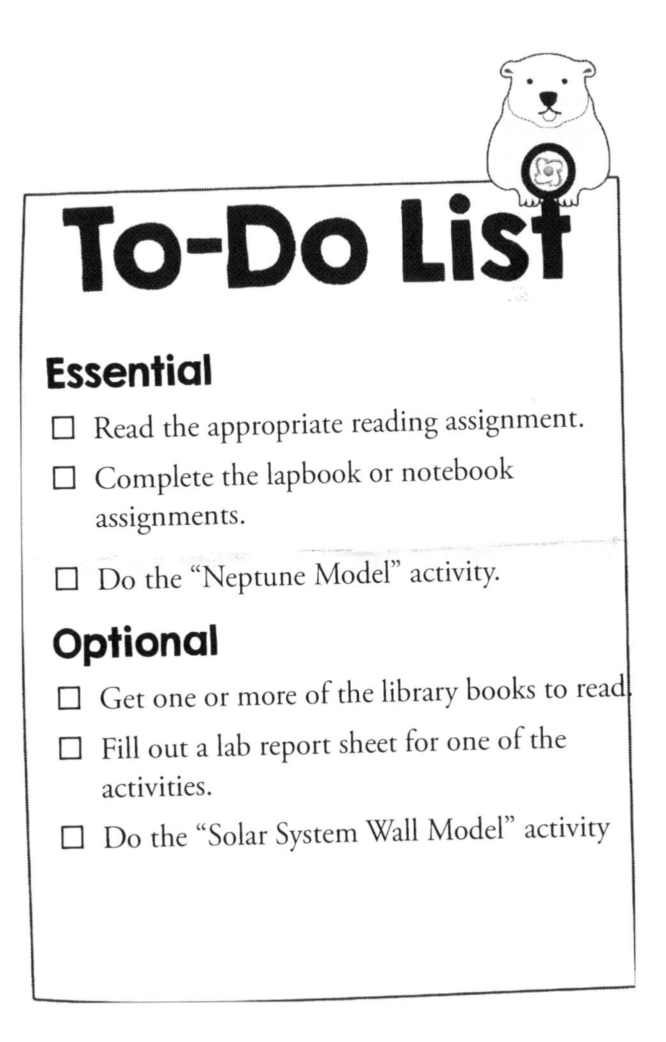

Lesson 11: Dwarf Planets

Information

Reading Assignments

- **Younger Students** – *DK Children's Encyclopedia* p. 196 Pluto
- **Older Students** – *Kingfisher Science Encyclopedia* pp. 410-411 The Solar System's Minor Members

(Optional) Books from the Library

- *Pluto: From Planet to Dwarf (True Books: Space)* by Elaine Landau
- *When Is a Planet Not a Planet?: The Story of Pluto* by Elaine Scott
- *Pluto: Dwarf Planet (Scholastic News Nonfiction Readers: Space Science)* by Christine Taylor-Butler
- *Pluto: The Dwarf Planet (Our Solar System)* by Greg Roza
- *The Planet Hunter: The Story Behind What Happened to Pluto* by Elizabeth Rusch and Guy Francis
- *Dwarf Planets: Pluto, Charon, Ceres, and Eris (Amazing Science: Planets)* by Nancy Loewen and Jeff Yesh

Notebooking

Vocabulary

Go over the following word with your students. Then, have them create a flashcard or copy the definition into the glossary.

- **Dwarf planet** – A celestial body that looks like a planet but does not meet the three requirements to be one. (Flashcard LT p. 67, Glossary NB p. 60)

Writing Instructions

- **Lapbook** – Have the students complete the Pluto tab-book on LT p. 60. Have them cut out the template and color the cover. Have them color in the location of the planet in our solar system and fill in the planet details. Then, have the students tell you what they have learned about Pluto and write it on the last page of the tab-book. Finally, have them glue the tab-book into the lapbook.

- **Notebook** – Have the students dictate, copy, or write two to six total sentences on dwarf planets on NB p. 36.

Hands-on Science

Coordinating Activities

- **Dwarf Planets** – Have the students research the known dwarf planets in our solar system.

Solar System Lesson Pages

Have them write a sentence or two about each. Their report should include information on Pluto, Eris, and Ceres. The following page from NASA will help your students begin their research:

https://solarsystem.nasa.gov/planets/profile.cfm?Object=Dwarf

(Optional) Solar System Wall Model – Pluto is technically a dwarf planet, but you can still add it to the solar system model. See the following chart for the size of the planet and its distance from the Sun.

Planet	Distance from the Sun	Scale diameter of planet
Pluto	13 feet, 3 inches	1/2 inches

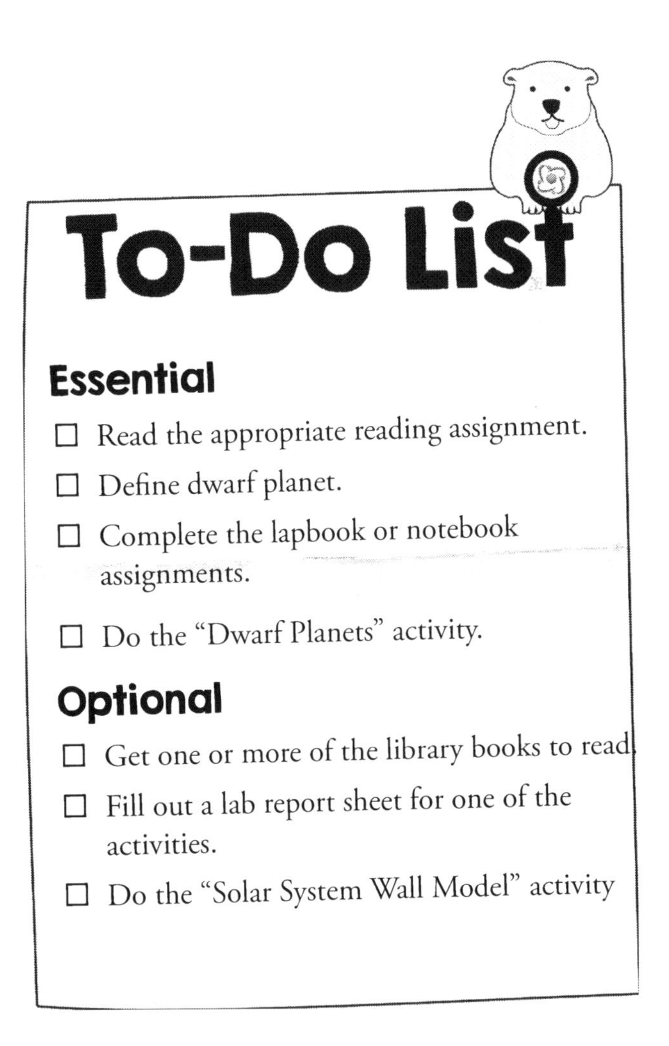

Lesson 12: Asteroids, Comets, and Meteors

Information

Reading Assignments

- **Younger Students** – *DK Children's Encyclopedia* p. 30 Asteroids, p. 68 Comets, p. 164 Meteorites
- **Older Students** – *Kingfisher Science Encyclopedia* p. 412 Comets, p. 413 Meteors and Meteorites

(Optional) Books from the Library

- *Comets, Meteors, and Asteroids* by Seymour Simon
- *Asteroids, Comets, and Meteorites (First Facts)* by Steve Kortenkamp
- *Asteroids and Comets (Science Readers: A Closer Look)* by William B. Rice
- *Comets and Asteroids: Space Rocks (Our Solar System)* by Greg Roza
- *Magic School Bus Out of This World : A Book about Space Rocks* by Joanna Cole and Bruce Degen

Notebooking

Vocabulary

Go over the following words with your students. Then, have them create a flashcard or copy the definition into the glossary.

- **Asteroid** – A rock orbiting the Sun. (Flashcard LT p. 67, Glossary NB p. 58)
- **Meteor** – A rock that travels through space and burns up when it enters a planet's atmosphere; also known as a shooting star. (Flashcard LT p. 68, Glossary NB p. 64)

Writing Instructions

- **Lapbook** – Have the students complete the Asteroids mini-book on LT p. 61. Have them cut out the template and fold along the dotted line. Have them tell you what they have learned about asteroids and meteors and write it on the inside of the mini-book. Then, have them color the picture and glue the mini-book into the lapbook.
- **Lapbook** – Have the students complete the Comets mini-book on LT p. 62. Have them cut out the template and fold along the dotted line. Have them tell you what they have learned about comets and write it on the inside of the mini-book. Then, have them color the picture and glue the mini-book into the lapbook.
- **Notebook** – Have the students dictate, copy, or write one to four sentences on asteroids, meteors, and comets on NB p. 37.

Hands-on Science

Coordinating Activities

- **Comets** – Have the students make their own comet. You will need about three feet of curling ribbon, a tennis ball, foil, and a straight pin. Have them cut the curling ribbon into lengths, curl them, and tie them together at one end. Next, have the students pin the tied end of the ribbon onto the tennis ball using the straight pin. Then, cut out a piece of foil, about 7" x 7", for them to use to cover the ball. Make sure they allow space for the comet's tail (ribbon) to come out through the foil. Once they are done, they can have their comets fly through outer space.
- **(Optional) Solar System Wall Model** – Have the students add an asteroid belt to their solar system wall model. You can have them add pictures of rocks or wadded-up brown paper in between Mars and Jupiter.

Review Sheet

Solar System Review Sheet Answers

1. Milky Way
2. True
3. Star
4. Gas
5. 1st
6. False *(A planet is large ball of rock or gas that travels around a star.)*
7. Sun
8. True
9. Water
10. Another Planet
11. Colder
12. True
13. Largest
14. False *(Jupiter's red spot is constantly moving.)*
15. Rocks and ice
16. Light
17. Gas
18. True
19. True

20. Toward
21. True
22. Dwarf planets
23. Rock
24. False *(Comets are large balls of rock and ice.)*
25. Shooting Stars

To-Do List

Essential

- ☐ Read the appropriate reading assignment.
- ☐ Define asteroid and meteor.
- ☐ Complete the lapbook or notebook assignments.
- ☐ Do the "Comets" activity.

Optional

- ☐ Get one or more of the library books to read.
- ☐ Fill out a lab report sheet for one of the activities.
- ☐ Do the "Solar System Wall Model" activity
- ☐ Complete the Solar System Review Sheet. (p. 129-131)

Atoms and Molecules Unit

Lesson 1: Atoms

Information

Reading Assignments

- **Younger Students** - *Basher Science Chemistry* p. 26 Atom, p. 28 Isotope
- **Older Students** - *Usborne Science Encyclopedia* p. 10-11 Atomic Structure, p. 13 Isotopes and Atomic Theory

(Optional) Books from the Library

- *What Are Atoms? (Rookie Read-About Science)* by Lisa Trumbauer
- *Atoms and Molecules (Building Blocks of Matter)* by Richard and Louise Spilsbury
- *Atoms (Simply Science)* by Melissa Stewart

Notebooking

Vocabulary

Go over the following words with your students. Then, have them create a flashcard or copy the definition into the glossary.

- **Electron** – A negatively charged particle in an atom. (Flashcard LT p. 77; Glossary NB p. 61)
- **Proton** – A positively charged particle in an atom. (Flashcard LT p. 78; Glossary NB p. 67)
- **Neutron** – A neutral particle in an atom.(Flashcard LT p. 78; Glossary NB p. 65)
- **Isotope** – An atom that has a different number of neutrons and so has a different mass number from the other atoms of an element. (Flashcard LT p. 79; Glossary NB p. 63)

Writing Instructions

- **Lapbook** – Have the students begin the Atoms lapbook by cutting out and coloring the cover on LT p. 71.
- **Lapbook** – Have the students complete the Atoms wheel-book on LT p. 72. Have them cut along the solid lines, punch a hole in the center, and use a brad fastener to fasten the two circles together. Have the students write their electron narration to the left of the picture, their proton narration above the picture, and their neutron narration to the right of the picture. Finally, have them glue their mini-book into the lapbook.
- **Lapbook** – Have the students complete the Isotopes shutterfold book on LT p. 73. Have them cut out and fold the template. Have the students color the pictures on the cover. Have them write their narration about the isotopes inside the mini-book. Then, have them glue the mini-book into the lapbook.
- **Notebook** – Have the students dictate, copy, or write one to four sentences about atoms on subatomic particles, atoms, and isotopes page on NB p. 40.

Hands-on Science

Coordinating Activities

- **Model Atom** – Have the students make an atom model. You will need 4 pipe cleaners and round beads in three different colors (at least 3 of each color). Have the students select which beads will be electrons, protons, and neutrons. Next, have them string three protons beads and three neutrons beads on one of the pipe cleaners, alternating between the two. Once done, have the students wrap this portion of the pipe cleaner into a ball to form a nucleus, leaving a straight end to connect to the electron rings they will make in the next step. Then, have the students place one electron bead on a pipe cleaner and twist the pipe cleaner closed to form a ring. Repeat this process two more times, so that they have 3 electron rings. Finally, fit the rings inside each other and then hang the nucleus ball in the center, using the pipe cleaner tail left in step two to attach the nucleus and hold the rings together. (*See image for reference.*)

- **(Optional) Subatomic Particles** – Make some subatomic cookies with your students using a sugar cookie, white icing, and three different colors of M&M's. See the following website for directions:
    - http://technoprairie.blogspot.com/2009/02/atomic-cookies.html

To-Do List

Essential

- ☐ Read the appropriate reading assignment.
- ☐ Define electron, proton, neutron, and isotope.
- ☐ Complete the lapbook or notebook assignments.
- ☐ Do the "Model Atom" activity.

Optional

- ☐ Get one or more of the library books to read.
- ☐ Fill out a lab report sheet (p. 101) for one of the activities.
- ☐ Do "Subatomic Particles" activity.

Lesson 2: Molecules

Information

Reading Assignments

- **Younger Students** - *Basher Science Chemistry* p. 32 Molecules
- **Older Students** - *Usborne Science Encyclopedia* p. 14-15 Molecules

(Optional) Books from the Library

- *Atoms and Molecules (Building Blocks of Matter)* by Richard and Louise Spilsbury
- *Atoms and Molecules (Why Chemistry Matters)* by Molly Aloian
- *Atoms and Molecules (My Science Library)* by Tracy Nelson Maurer

Notebooking

Vocabulary

Go over the following words with your students. Then, have them create a flashcard or copy the definition into the glossary.

- **Electron shell** – The region around an atom's nucleus in which a certain amount of electrons can reside. (Flashcard LT p. 79; Glossary NB p. 61)
- **Molecule** – A substance made up of two or more atoms that are chemically bonded. (Flashcard LT p. 80; Glossary NB p. 65)

Writing Instructions

- **Lapbook** – Have the students work on the Electron Shell Diagram on LT p. 73. Have the students cut out the sheet, color the shells different colors, and add the information they have learned about how many electrons the first three shells can carry. Finally, have them glue their sheets into their lapbooks.

- **Lapbook** – Have the students work on the Molecules tab-book on LT p. 74. Have the students write the definition of a molecule on the definition page and then add any molecules they have learned about to the samples page. Set the mini-book aside and save it for next week.
- **Notebook** – Have the students dictate, copy, or write one to four sentences on electron shells, molecules, and nonpolar and polar molecules on NB p. 41.

Hands-on Science

Coordinating Activities

- **Electron Shells** – Have the students play the atoms and isotopes game, focusing on reviewing how many electrons are in each shell. You can get directions for this game from the following blog post:

- http://elementalscience.com/blogs/science-activities/60317571-free-chemistry-game

(Optional) Molecules – Have the students make molecules models out of LEGOS using the examples from the following pin:

- https://www.pinterest.com/pin/192036371586132562/

To-Do List

Essential

- ☐ Read the appropriate reading assignment.
- ☐ Define electron shell and molecule.
- ☐ Complete the lapbook or notebook assignments.
- ☐ Do the "Electron Shells" activity.

Optional

- ☐ Get one or more of the library books to read.
- ☐ Fill out a lab report sheet for one of the activities.
- ☐ Do the "Molecules" activity.

Lesson 3: Air

Information

Reading Assignments

- **Younger Students** - *Basher Science Chemistry* p. 96 Air, p. 110 Oxygen, p. 112 Carbon Dioxide
- **Older Students** - *Usborne Science Encyclopedia* p. 62-63 Air

(Optional) Books from the Library

- *Air Is All Around You (Let's-Read-and-Find... Science 1)* by Franklyn M. Branley
- *Air: Outside, Inside, and All Around (Amazing Science)* by Darlene R. Stille

Notebooking

Vocabulary

Go over the following words with your students. Then, have them create a flashcard or copy the definition into the glossary.

- **Air** – A mixture of gases that form a protective layer around the Earth. (Flashcard LT p. 80; Glossary NB p. 58)

Writing Instructions

- **Lapbook** – Have the students complete the Air mini-book on LT p. 75. Have them cut out and fold the template. Have the students color the pictures on the cover. Have them write their narration about the air inside the mini-book. Then, have them glue the mini-book into the lapbook.
- **Notebook** – Have the students dictate, copy, or write one to four sentences on air, oxygen, and carbon dioxide on NB p. 42.

Hands-on Science

Coordinating Activities

- **Air** – Have the students play a game with air. You will need a balloon for this activity. Blow up the balloon, sharing with the students that air is what fills the balloons. Then, hit the balloon back and forth to each other. The goal of the game is to keep the balloon from touching the ground. See how many times you can go back and forth without doing so!
- **(Optional) Burning Oxygen** – Have the students see how oxygen is need for combustion. You will need a candle and a clear glass bottle for this activity. Light the candle and let it burn for a bit. Then, place the glass bottle over the candle and watch what happens. (*The candle will*

burn for a bit before going out. This is because it uses up all the oxygen trap ed in the air in the bottle.)

To-Do List

Essential

- ☐ Read the appropriate reading assignment.
- ☐ Define air.
- ☐ Complete the lapbook or notebook assignments.
- ☐ Do the "Air" activity.

Optional

- ☐ Get one or more of the library books to read.
- ☐ Fill out a lab report sheet for one of the activities.
- ☐ Do the "Burning Oxygen" activity.

Lesson 4: Water

Information

Reading Assignments

- **Younger Students** - *Basher Science Chemistry* p. 108 Water
- **Older Students** - *Usborne Science Encyclopedia* p. 72-73 Water

(Optional) Books from the Library

- *Water, Water Everywhere (Reading Rainbow Book)* by Cynthia Overbeck Bix
- *Water* by Frank Asch
- *Water: Up, Down, and All Around (Amazing Science)* by Natalie M. Rosinsky

Notebooking

Vocabulary

Go over the following words with your students. Then, have them create a flashcard or copy the definition into the glossary.

- **Hard water** – Water which contains a lot of dissolved minerals. (Flashcard LT p. 81; Glossary NB p. 62)

Writing Instructions

- **Lapbook** – Have the students complete the Water mini-book on LT p. 76. Have them cut out and fold the template. Have the students color the pictures on the cover. Have them write their narration about water inside the mini-book. Then, glue the mini-book into the lapbook.
- **Notebook** – Have the students dictate, copy, or write one to four sentences on water, water as a solvent, and hard water on NB p. 43.

Hands-on Science

Coordinating Activities

- **Floating Water** – Test whether ice is less dense than water. You will need a cup and several cubes of ice. Fill the cup two-thirds of the way full with water. Add two to three ice cubes and observe what happens.
- **(Optional) Water Art** – Have the students paint with water colors! As they create their pictures, discuss the fact that they are able to paint with the colors because water is such a good solvent.

Review Sheet

Atoms and Molecules Review Sheet Answers

1. Positive, Negative, Neutral
2. True
3. 2, 8, 8 to 18
4. False (*A molecule can be made up of more than one element.*)
5. Nitrogen, Oxygen
6. Oxygen, Carbon dioxide, Carbon dioxide, Oxygen
7. More, Less
8. True

To-Do List

Essential

- ☐ Read the appropriate reading assignment.
- ☐ Define hard water.
- ☐ Complete the lapbook or notebook assignments.
- ☐ Do the "Floating Water" activity.

Optional

- ☐ Get one or more of the library books to read.
- ☐ Fill out a lab report sheet for one of the activities.
- ☐ Do the "Water Art" activity.
- ☐ Complete the Atoms and Molecules Review Sheet. (p. 132)

Light Unit

Lesson 1: Light

Information

Reading Assignments

- **Younger Students** - *DK Children's Encyclopedia* p. 147 Light
- **Older Students** - *Usborne Science Encyclopedia* pp. 214-215 Light and Shadow

(Optional) Books from the Library

- *Light Is All Around Us (Let's-Read-and-Find-Out Science 2)* by Wendy Pfeffer and Paul Meisel
- *All About Light (Rookie Read-About Science)* by Lisa Trumbauer
- *Day Light, Night Light: Where Light Comes From (Let's-Read-and-Find-Out Science 2)* by Dr. Franklyn M. Branley and Stacey Schuett

Notebooking

Vocabulary

Go over the following words with your students. Then, have them create flashcards or copy the definitions into the glossary.

- **Light** – The electromagnetic waves of energy that make objects visible. (Flashcard LT p. 91; Glossary NB p. 64)
- **Shadow** – A dark area that is formed when an object blocks out light waves. (Flashcard LT p. 91; Glossary NB p. 69)

Writing Instructions

- **Lapbook** – Have the students begin the Light lapbook by cutting out and coloring the cover on LT p. 84. Then, have the students glue the sheet onto the front.
- **Lapbook** – Have the students cut out and glue the vocabulary pocket on LT p. 90 into their lapbook.
- **Lapbook** – Have the students complete the Light versus Shadow Shutterfold book on LT p. 86. Have them cut out and fold the template. Have the students color the pictures on the cover. Then, have the students write the definition of light under the light flap and the definition of shadow under the shadow flap. Finally, have them glue the mini-book into the lapbook.
- **Lapbook** – Have the students add the "Light" poem to the lapbook. Have them cut out and color the poem sheet found on LT p. 85. Once they are done, have them glue the sheet into the lapbook. (**Note** - *You can have the students memorize this poem as you work through this unit.*)
- **Notebook** – Have the students dictate, copy, or write three to five sentences about light on

the light notebooking page on NB p. 46.

Hands-on Science

Coordinating Activities

- **Sight Box** – You will need a small nail or screw, a box with a lid, several small objects (such as a ball, a pencil, or a toy car), and a flashlight. Use a small nail or screw to make a pinhole at the end of one side of the box. Place the small objects inside the box, and then close the lid tightly. Ask the students to look inside the hole to see what is in the box. Then, have the students step back as you turn on the flashlight. Place the flashlight in the box opposite from the objects. Ask the students to look inside the hole again to see what is in the box. (*The students should not be able to see the objects when the cover is on, and they should be able to see the objects when the flashlight is on. We see objects because light bounces off the objects and is reflected back to our eyes, letting our brain know that something is there. If there is no source of light, we cannot see the objects that are there.*)
- **(Optional) Light Camera** – Have the students make a pinhole camera. You will need a round oatmeal container, tissue paper, aluminum foil, a pin, a knife, tape, black construction paper, and a flashlight. Remove the top of the oatmeal container, cover it with tissue paper, and use tape to secure it in place. Then, flip it upside down and cut a small square in the bottom of the container. Cover this hole with foil, and use tape to secure it in place. Next, use the pin to make a small hole in the center of the foil. After that, cut a small shape, such as a triangle or small paper doll, out of the construction paper. Now, take all that you have made into a darkened room. Set up the box about two feet from the flashlight so that the flashlight is pointed toward the pinhole in the bottom of the oatmeal container. Turn the flashlight on, and put the black construction paper shape in front of the flashlight, about an inch or so away from the flashlight. Then, have the students look at the tissue paper covering the top of the oatmeal container to observe what they see. (*The students should see the image of their shape upside down. This is because the light travels straight from the top and bottom of the object at an angle and causes the image to be flipped, like it does in our eyes, and then our brain turns the image right side up again.*)

To-Do List

Essential

- ☐ Read the appropriate reading assignment.
- ☐ Define light and shadow.
- ☐ Complete the lapbook or notebook assignments.
- ☐ Do the "Sight Box" activity.

Optional

- ☐ Get one or more of the library books to read.
- ☐ Fill out a lab report sheet (p. 101) for one of the activities.
- ☐ Do the "Light Camera" activity.

Lesson 2: Color

Information

Reading Assignments

- **Younger Students** - *DK Children's Encyclopedia* p. 26-27 Color
- **Older Students** - *Usborne Science Encyclopedia* pp. 216-217 Color

(Optional) Books from the Library

- *Pantone: Colors* by Pantone and Helen Dardik
- *All the Colors of the Rainbow (Rookie Read-About Science)* by Allan Fowler
- *Color Day Relay (The Magic School Bus Chapter Book)* by Gail Herman and Hope Gangloff

Notebooking

Vocabulary

Go over the following words with your students. Then, have them create flashcards or copy the definitions into the glossary.

- **Primary colors** – Colors from which all other colors can be made. The primary colors are red, yellow, and blue. (Flashcard LT p. 92; Glossary NB p. 66)
- **Secondary colors** – Colors that can be made by mixing two primary colors. The secondary colors are orange, green, and purple. (Flashcard LT p. 92; Glossary NB p. 68)

Writing Instructions

- **Lapbook** – Have the students complete the Colors Tab-book on LT p. 87. Have them cut out the pages for the tab-book and color the pictures. Then, have the students add a sentence about primary colors, along with the colors that are considered primary colors, on the primary page. Have them add a sentence about secondary colors, along with the colors that are considered secondary colors, on the secondary page. Assemble the tab-book and staple it together on the dashed lines. Finally, have the students glue the mini-book into the lapbook.
- **Notebook** – Have the students dictate, copy, or write three to five sentences on the color notebooking page on NB p. 47.

Hands-on Science

Coordinating Activities

- **Color Mixing** – Have the students create a color wheel by mixing paint. You will need a piece of paper, paint (red, yellow, and blue), pencil, and a paintbrush. Draw a circle and divide it into six parts. Begin with the primary colors by painting one part red and skip the next section. Repeat this process for both the yellow and blue. Then, add the secondary colors by mixing the two surrounding primary colors in the blank section, i.e., if the blank section is in between red and yellow, mix red and yellow to make orange. Repeat the process until the wheel has all the primary and secondary colors.

✂ **(Optional) Up-close Color** – Have the students look at a printed color cover or page up close, using a magnifying glass. (**Note** - *You can get better results for this activity with either a currency detection scope or a microscope.*)

To-Do List

Essential

- ☐ Read the appropriate reading assignment.
- ☐ Define primary and secondary colors.
- ☐ Complete the lapbook or notebook assignments.
- ☐ Do the "Color Mixing" activity.

Optional

- ☐ Get one or more of the library books to read.
- ☐ Fill out a lab report sheet for one of the activities.
- ☐ Do the "Up-close Color" activity.

Lesson 3: Light Behavior

Information

Reading Assignments

- **Younger Students** - No pages are scheduled for younger students. Instead, have the students watch the following video to learn more about light behavior:
  - https://www.youtube.com/watch?v=JRh0CGfX7dQ
- **Older Students** - *Usborne Science Encyclopedia* pp. 218-219 Light Behavior

(Optional) Books from the Library

- *Shadows and Reflections (Light All Around Us)* by Daniel Nunn
- *Shadows and Reflections* by Tana Hoban
- *What Are Shadows and Reflections? (Light & Sound Waves Close-Up)* by Robin Johnson

Notebooking

Vocabulary

Go over the following words with your students. Then, have them create flashcards or copy the definitions into the glossary.

- **Reflection** – The change in direction of light rays that occurs when it hits an object and bounces off. (Flashcard LT p. 93; Glossary NB p. 67)
- **Refraction** – The bending of light rays caused by light passing through substances with different densities. (Flashcard LT p. 93; Glossary NB p. 67)

Writing Instructions

- **Lapbook** – Have the students complete the Light Behavior Triangle Book on LT p. 88. Have them cut out the pages for the triangle book and color the pictures. Then, have the students add a sentence about reflection, refraction, and diffraction on the respective flaps. Assemble the triangle book and glue on the cover for the mini-book. Finally, have the students glue the mini-book into the lapbook.

- **Notebook** – Have the students dictate, copy, or write three to five sentences on light behavior on the light behavior notebooking page on NB p. 48.

Hands-on Science

Coordinating Activities

- **Reflection** – Have the students use a homemade kaleidoscope to see reflection in action. You will need an empty toilet paper roll, a thick mylar sheet, scissors, tape, card stock, a straw, and markers. Find the directions for this activity here:
  - https://buggyandbuddy.com/science-for-kids-how-to-make-a-kaleidoscope/

✂ **(Optional) Refraction** – Have the students do the "See for yourself" activity on p. 218 of the *Usborne Science Encyclopedia*. You will need a glass of water and a straw for this activity.

To-Do List

Essential

- ☐ Read the appropriate reading assignment.
- ☐ Define reflection and refraction.
- ☐ Complete the lapbook or notebook assignments.
- ☐ Do the "Reflection" activity.

Optional

- ☐ Get one or more of the library books to read.
- ☐ Fill out a lab report sheet for one of the activities.
- ☐ Do the "Refraction" activity.

Lesson 4: Lenses and Mirrors

Information

Reading Assignments

- **Younger Students -** No pages are scheduled for younger students. Instead, have the students watch the following videos to learn more about lens and mirrors:
    - Convex lens - https://www.youtube.com/watch?v=cf_aUBbyuts
    - Mirror - https://www.youtube.com/watch?v=N6n0FAZ_6N8
- **Older Students -** *Usborne Science Encyclopedia* pp. 220-221 Lens and Mirrors

(Optional) Books from the Library

- *Light: Shadows, Mirrors, and Rainbows (Amazing Science)* by Natalie M. Rosinsky and Sheree Boyd

Notebooking

Vocabulary

Go over the following words with your students. Then, have them create flashcards or copy the definitions into the glossary.

- **Lens** – A curved transparent surface that causes light to bend in a particular way. (Flashcard LT p. 94; Glossary NB p. 64)
- **Mirror** – A shiny surface that reflects nearly all the light that hits it. (Flashcard LT p. 94; Glossary NB p. 65)

Writing Instructions

- **Lapbook –** Have the students complete the Lenses and Mirrors Mini-book on LT p. 89. Have them cut out and fold the template. Have the students color the pictures on the cover. Then, have the students write the difference between lenses and mirrors on the inside. Finally, have them glue the mini-book into the lapbook.
- **Notebook –** Have the students dictate, copy, or write three to five sentences on lenses and mirrors on the lens and mirrors notebooking page on NB p. 49.

Hands-on Science

Coordinating Activities

- **Glass Lens –** Have the students make their own reading "glasses." You will need a glass jar, water, pencil, and an index card. Have the students fill the glass jar with water as you write a message on the index card in very small letters. Then, have the students hold the index card up to the jar so that they are looking through the jar to read the card. Have them read your message aloud. (*They should see that the message appears larger as the curved glass jar and water serve as a magnifying glass.*)

Review Sheet

Light Review Sheet Answers

1. True
2. Opaque
3. Red, Orange, Yellow, Green, Blue, Indigo, Violet
4. False *(Different colors have different wavelengths and frequencies.)*
5. When light is reflected, the rays hit an object and bounce off it.
6. When light is refracted, the rays are bent as they pass from one substance through another one of differing density. (**Note** - *If your students have only noted that the rays are bent, you can mark this answer correct.)*
7. False *(A lens is a curved transparent surface that causes light to bend in a particular way.)*
8. False *(A mirror is a shiny surface that reflects nearly all the light that hits it.)*

To-Do List

Essential

- ☐ Read the appropriate reading assignment.
- ☐ Define lens and mirror.
- ☐ Complete the lapbook or notebook assignments.
- ☐ Do the "Glass Lens" activity.

Optional

- ☐ Get one or more of the library books to read.
- ☐ Fill out a lab report sheet for one of the activities.
- ☐ Complete the Light Review Sheet. (pp. 133-134)

Sound Unit

Lesson 1: Sound

Information

Reading Assignments

- **Younger Students** - *DK Children's Encyclopedia* p. 235 Sound
- **Older Students** - *Usborne Science Encyclopedia* pp. 206-207 Sound

(Optional) Books from the Library

- *Sound: Loud, Soft, High, and Low (Amazing Science)* by Natalie M. Rosinsky and Matthew John
- *Sounds All Around (Let's-Read-and-Find-Out Science 1)* by Wendy Pfeffer and Anna Chernyshova
- *All About Sound (Rookie Read-About Science)* by Lisa Trumbauer

Notebooking

Vocabulary

Go over the following words with your students. Then, have them create a flashcard or copy the definition into the glossary.

- **Decibel (dB)** – The unit of loudness. (Flashcard LT p. 105; Glossary NB p. 59)
- **Sound wave** – A mechanical wave that carries sound energy through a medium. (Flashcard LT p. 105; Glossary NB p. 69)

Writing Instructions

- **Lapbook** – Have the students begin the Sound lapbook by cutting out and coloring the cover on LT p. 97. Then, have the students glue the sheet onto the front.
- **Lapbook** – Have the students cut out and glue the vocabulary pocket on LT p. 104 into their lapbook.
- **Lapbook** – Have the students complete the Sound Mini-book on LT p. 98. Have the students color the pictures on the cover. Then, have the students write the definition of sound inside the mini-book. Finally, have them glue the mini-book into the lapbook.
- **Lapbook** – Have the students add the "Sound" poem to the lapbook. Have them cut out and color the poem sheet found on LT p. 99. Once they are done, have them glue the sheet into the lapbook. (**Note** - *You can have the students memorize this poem as you work through this unit.*)
- **Notebook** – Have the students dictate, copy, or write three to five sentences about sound on the sound notebooking page on NB p. 52.

Hands-on Science

Coordinating Activities

- **Sound** – Have the students feel the vibrations that sound makes. Have them place a hand lightly on their neck and then begin to hum or sing. (*The students should feel the vibrations in their throat. This is because our vocal cords vibrate to produce sound that can be heard.*)
- **(Optional) Echo Game** – Have the students play a game of echolocation, otherwise known as Marco Polo. Choose one person to be "it" and blindfold that person. (*Be sure that you are playing this game in a location free from trip-hazards.*) To begin the game, the blindfolded player cries out "Marco" and the other players respond with "Polo." The blindfolded player then tries to tag another player based on where he or she heard them. When the blindfolded player finally tags someone, that person becomes the next blindfolded Marco.

To-Do List

Essential

- ☐ Read the appropriate reading assignment.
- ☐ Define decibel and sound wave.
- ☐ Complete the lapbook or notebook assignments.
- ☐ Do the "Sound" activity.

Optional

- ☐ Get one or more of the library books to read.
- ☐ Fill out a lab report sheet for one of the activities.
- ☐ Do the "Echo Game" activity.

Lesson 2: Waves

Information

Reading Assignments

- **Younger Students** - *No pages are scheduled for younger students. Instead, have the students watch the following video to learn more about the two types of waves:*
  - https://www.youtube.com/watch?v=RVyHkV3wIyk
- **Older Students** - *Usborne Science Encyclopedia* pp. 202-203 Waves

(Optional) Books from the Library

- *Sound Waves and Communication (Science Readers)* by Jenna Winterberg
- *What Are Sound Waves? (Light & Sound Waves Close-Up)* by Robin Johnson
- *Sound Waves (Energy in Action)* by Ian F. Mahaney
- *The Science of Sound Waves (Catch a Wave)* by Robin Johnson

Notebooking

Vocabulary

Go over the following words with your students. Then, have them create a flashcard or copy the definition into the glossary.

- **Longitudinal wave** – A wave that vibrates in the same direction as it travels. (Flashcard LT p. 106; Glossary NB p. 64)
- **Transverse wave** – A wave that vibrates at right angles to the direction of travel. (Flashcard LT p. 106; Glossary NB p. 70)

Writing Instructions

- **Lapbook** – Have the students complete the Waves Shutterfold-book on LT p. 100. Have them cut out the mini-book template and color the pictures. Then, have the students add a sentence about longitudinal waves under the picture of the longitudinal wave (upper) and a sentence about transverse waves under the picture of the transverse wave (lower.) Finally, fold the mini-book along dashed lines and have the students glue the mini-book into the lapbook.
- **Notebook** – Have the students dictate, copy, or write three to five sentences on waves on the waves notebooking page on NB p. 53.

Hands-on Science

Coordinating Activities

- **Water Waves** – Have the students make waves in water. You will need a bowl of water and a small pebble. Have the students gently drop the pebble into the center of the bowl. Have them observe the resulting waves that are created.

✂ **(Optional) String Waves** – Have the students do the "See for yourself " activity on p. 203 of the *Usborne Science Encyclopedia.* You will need a string for this activity

To-Do List

Essential

- ☐ Read the appropriate reading assignment.
- ☐ Define longitudinal and transverse waves.
- ☐ Complete the lapbook or notebook assignments.
- ☐ Do the "Water Waves" activity.

Optional

- ☐ Get one or more of the library books to read.
- ☐ Fill out a lab report sheet for one of the activities.
- ☐ Do the "String Waves" activity.

Lesson 3: Wave Behavior

Information

Reading Assignments

- **Younger Students** - *No pages are scheduled for younger students. Instead, have the students watch the following video to learn more about wave behavior:*
  - https://www.youtube.com/watch?v=TfYCnOvNnFU
- **Older Students** - *Usborne Science Encyclopedia* pp. 204-205 Wave Behavior

(Optional) Books from the Library

- *How Does Sound Change? (Sound Waves Close-Up)* by Robin Johnson
- *How Sound Moves (Science Readers: Content and Literacy)* by Sharon Coan
- *Waves and Information Transfer (Catch a Wave)* by Heather C Hudak

Notebooking

Vocabulary

Go over the following words with your students. Then, have them create a flashcard or copy the definition into the glossary.

- **Interference** – The effect that occurs when two waves meet. (Flashcard LT p. 107; Glossary NB p. 63)

Writing Instructions

- **Lapbook** – Have the students complete the Wave Behavior Pocket Guide on LT p. 101-102. Have them cut out the pages for the pocket and the cards. Then, have the students color the pictures. If you are using the blank cards, have the students add a sentence about how waves are affected by reflection, refraction, interference, and diffraction on the respective cards. Finally, have the students glue the pocket into the lapbook and insert the cards.
- **Notebook** – Have the students dictate, copy, or write three to five sentences on wave behavior on the wave behavior notebooking page on NB p. 54.

Hands-on Science

Coordinating Activities

- **Giant Wave** – Make a giant wave for your students to play with. You will need string, popsicle sticks, ruler, pencil, and a hot glue gun. The directions for this project can be found here:
  - http://blog.teachersource.com/2015/01/13/making-waves/
- **(Optional) Interference** – Have the students do the "See for yourself" activity on p. 205 of the *Usborne Science Encyclopedia*. You will need two small pebbles and a bathtub full of water.

To-Do List

Essential

- ☐ Read the appropriate reading assignment.
- ☐ Define interference.
- ☐ Complete the lapbook or notebook assignments.
- ☐ Do the "Giant Wave" activity.

Optional

- ☐ Get one or more of the library books to read.
- ☐ Fill out a lab report sheet for one of the activities.
- ☐ Do the "Interference" activity.

Lesson 4: Musical Instruments

Information

Reading Assignments

- **Younger Students** - *DK Children's Encyclopedia* p. 176-177 Music
- **Older Students** - *Usborne Science Encyclopedia* pp. 208-209 Musical Instruments

(Optional) Books from the Library

- *The Science of Music (Super-Awesome Science)* by Cecilia Pinto McCarthy
- *Science of Music: Discovering Sound (Science in Action)* by Karen Latchana Kenney
- *Musical Instruments (How Things Work)* by Ade Deane-pratt

Notebooking

Vocabulary

Go over the following words with your students. Then, have them create a flashcard or copy the definition into the glossary.

- **Resonate** – To vibrate at the same frequency as something else. (Flashcard LT p. 107; Glossary NB p. 68)

Writing Instructions

- **Lapbook** – Have the students complete the How Instruments Work Tab-book on LT p. 103. Have them cut out the pages for the tab-book and color the pictures. Then, have the students add a sentence about how stringed instruments work on the string page, how wind instruments work on the wind page, and how percussion instruments work on the percussion page. Assemble the tab-book and staple it together on the dashed lines. Finally, have the students glue the mini-book into the lapbook.
- **Notebook** – Have the students dictate, copy, or write three to five sentences on musical instruments on the musical instruments notebooking page on NB p. 55.

Hands-on Science

Coordinating Activities

- **Physics of Music** – Have the students listen to some classical music. You can find directions for this activity here:
  - https://elementalscience.com/blogs/science-activities/physics-of-music

Review Sheet

Sound Review Sheet Answers

1. True
2. False *(An echo of a sound wave can be used to determine position.)*
3. Waves carry energy.
4. True
5. B, A
6. False *(A wave can change speed, direction, or shape when it passes into a different medium.)*
7. Interference
8. A, C, B

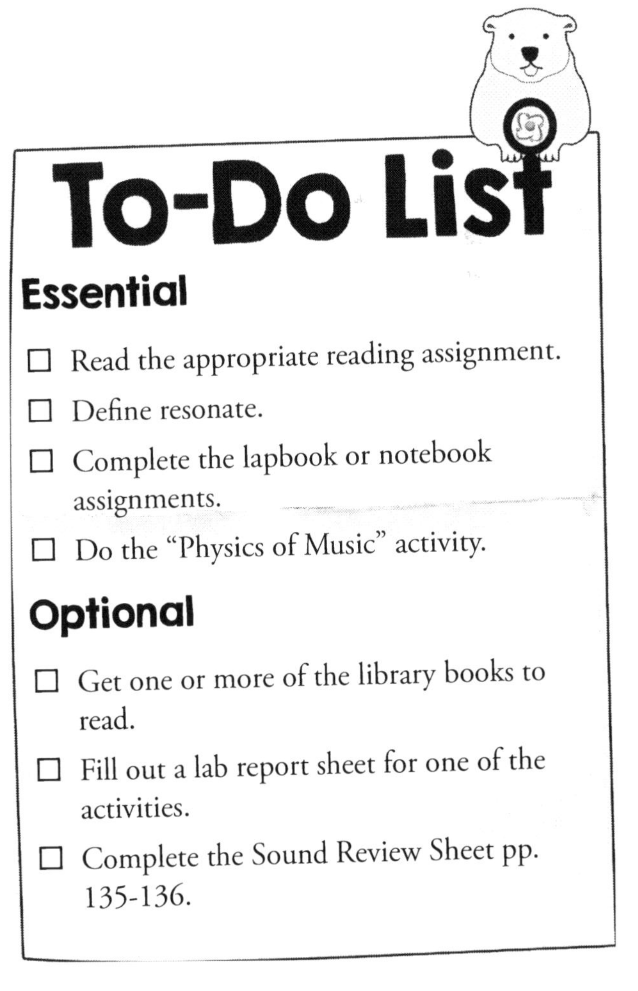

Appendix

Activity: ______________________________

What I learned

Lab Report: ______________________________

Our Tools

______________________ ______________________

______________________ ______________________

______________________ ______________________

Our Method

What it looked like

____

____

____

____

____

____

____

____

____

Our Outcome

____

____

____

____

____

Our Insight

____

____

____

____

Plant Growth Chart

18						
17						
16						
15						
14						
13						
12						
11						
10						
9						
8						
7						
6						
5						
4						
3						
2						
1						
Inches	Week 1	Week 2	Week 3	Week 4	Week 5	Week 6

Types of Roots

Roots are important to plants. They help to anchor the plant to the ground and to take in the water and nutrients the plant needs. As the roots of a plant grow, they will spread out as far as they can to reach as much water as possible.

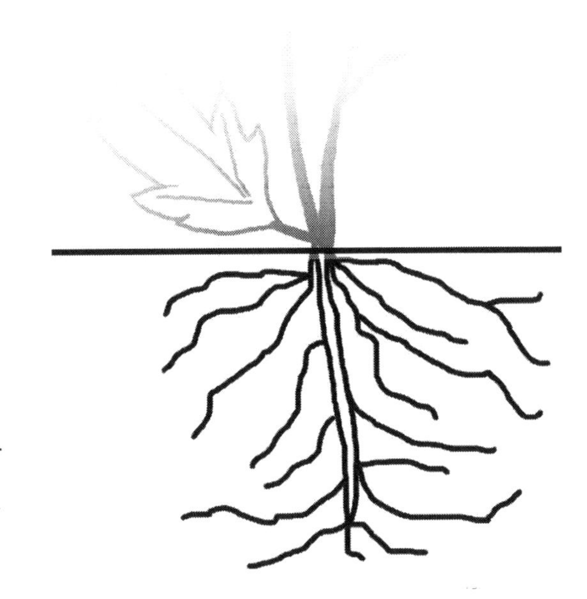

Plants have two basic types of roots – fibrous and taproots. These are classified by how the roots develop and branch out.

Fibrous roots are the most common root types. These roots are all about the same size, meaning that there is not a large single root in the center. Fibrous roots branch out multiple times as they grow. Plants with fibrous roots will have a mass of similarly sized roots at their base. The grass in your yard is a good example of a plant with fibrous roots.

Taproots, on the other hand, develop a single central root that is faster and deeper than the other branches. Plants with taproots have a large primary root with smaller rootlets that develop off of the main root. The dandelion in your yard is a good example of a plant with a taproot.

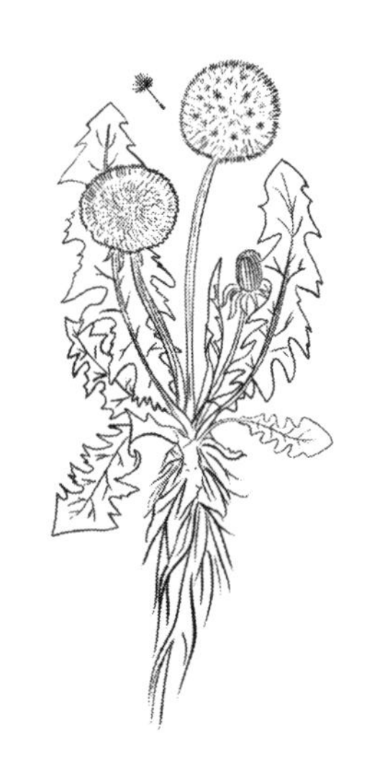

Facts about the Polar Biome

Facts about the Forest Biome

Facts about the Grasslands Biome

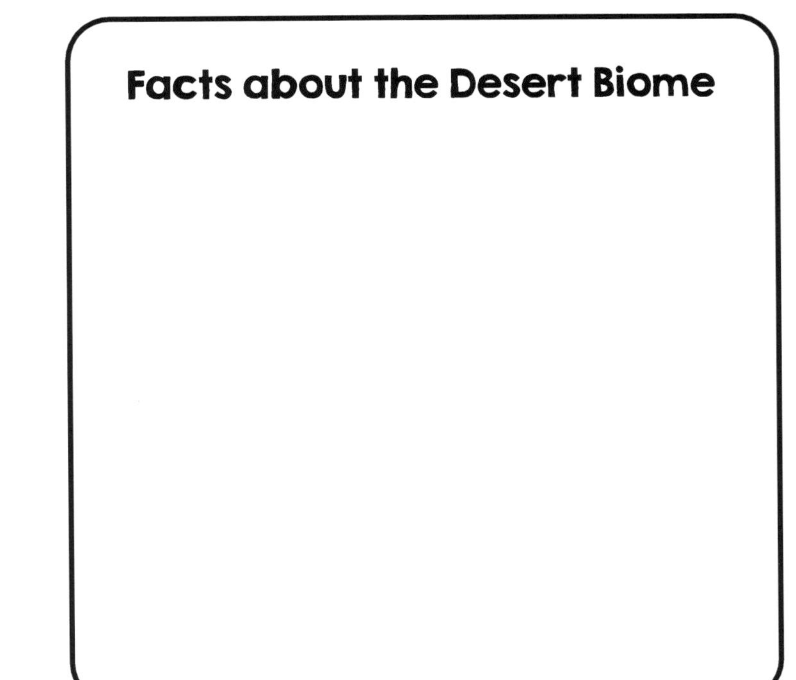

Facts about the Desert Biome

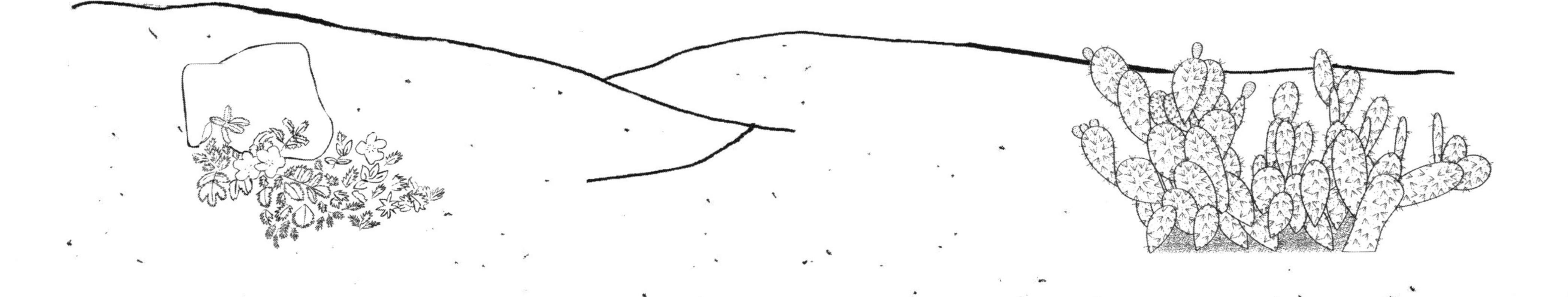

Facts about the Aquatic Biome

Tropical versus Temperate

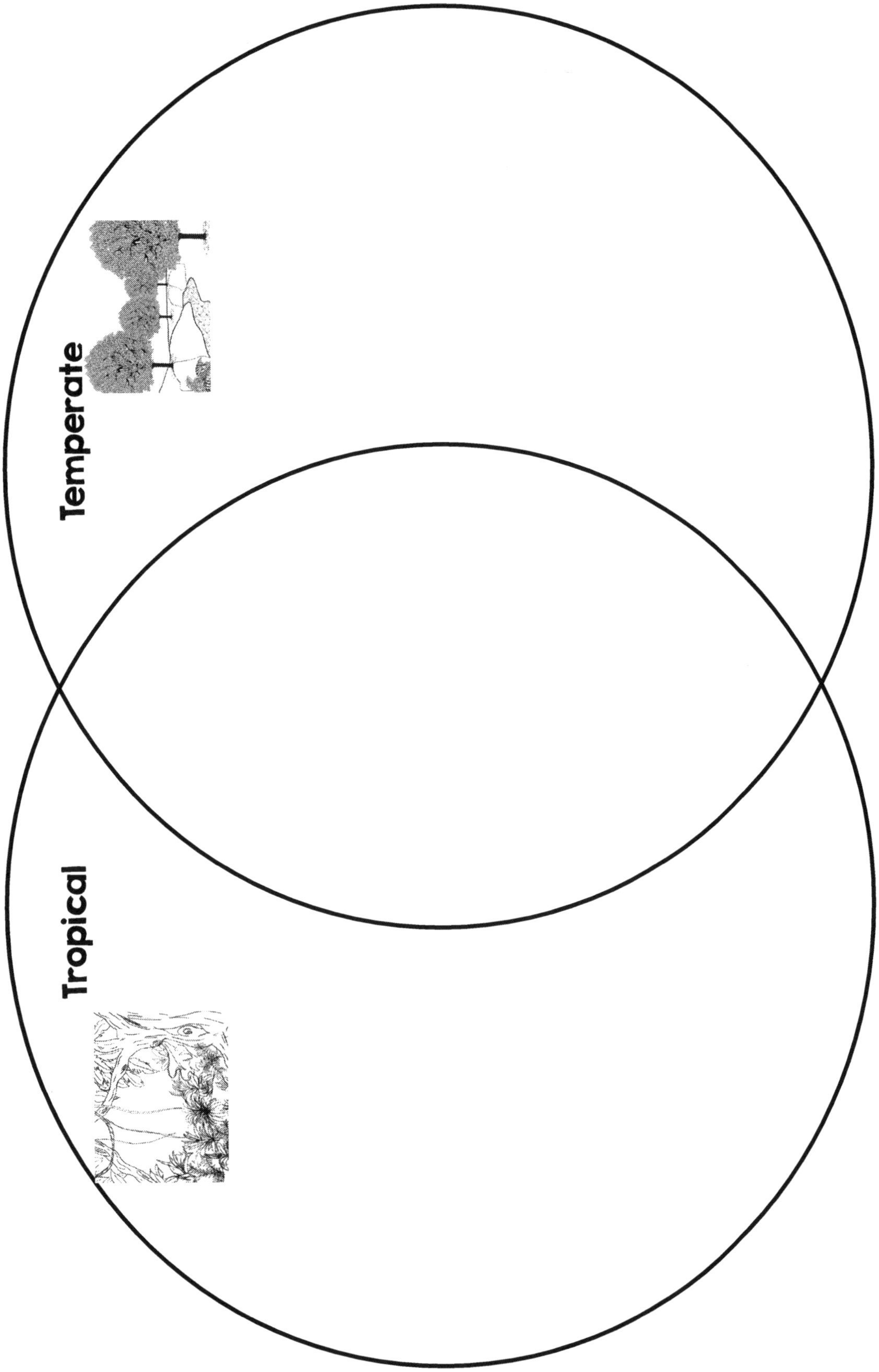

Temperate

Tropical

Planet Templates for Projects

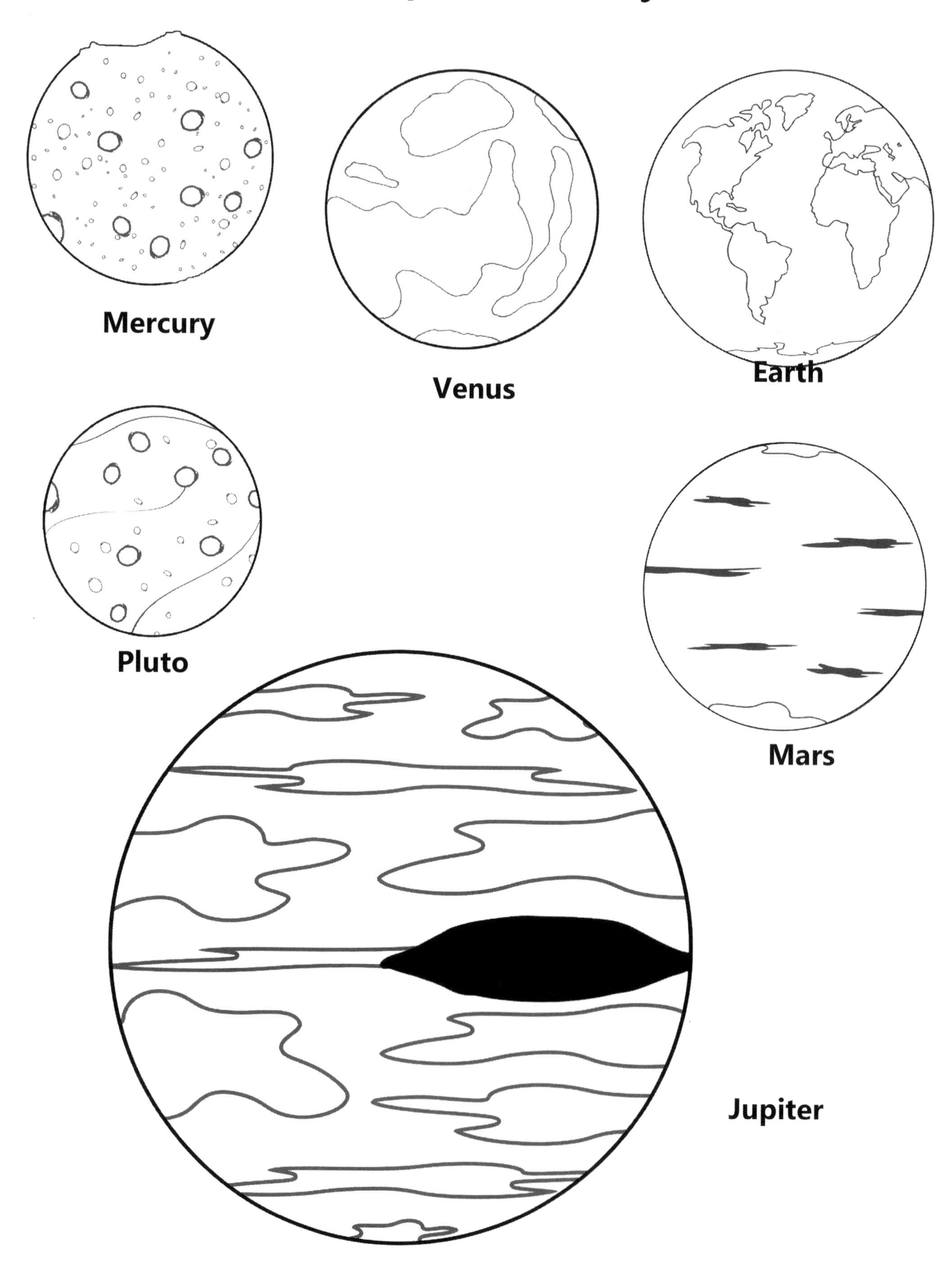

Planet Templates for Projects

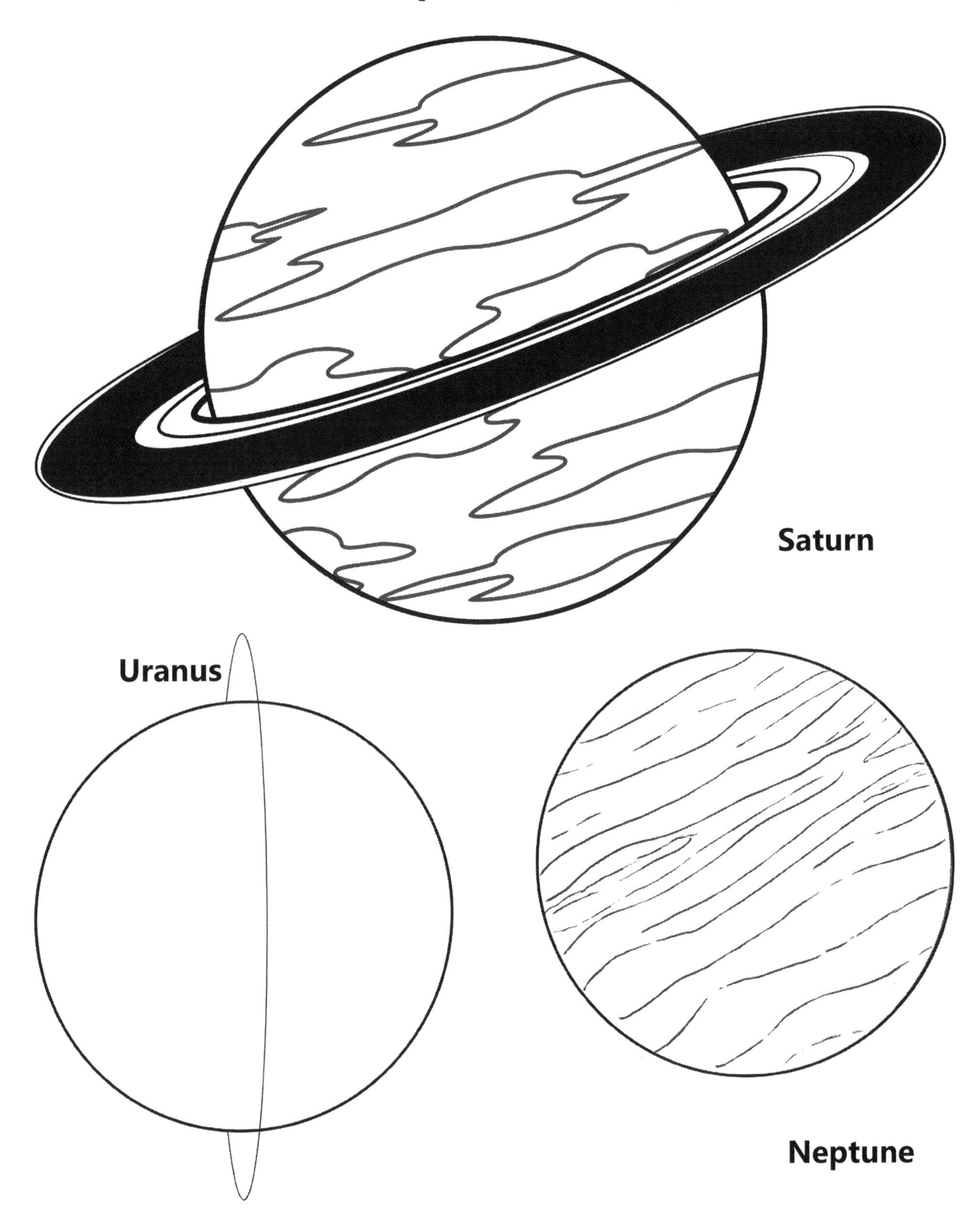

Paper-mâché Planet Directions

Materials Need

- ✂ Balloon
- ✂ Newspaper
- ✂ 1 Cup of flour
- ✂ ½ Cup of water
- ✂ 2 Tbsp of salt
- ✂ Picture of the planet

Steps to Complete

1. Begin by having the students blow up the balloon.
2. Next, have them tear the newspaper into strips. As they are working on the newspaper strips, use the flour, water, and salt to make a thick paste. You can add more or less water to gain the desired consistency.
3. Then, have the students dip the strips into the paste mixture and cover the balloon with one layer.
4. Wait 30 minutes before having them add a second layer. As they do this, have them look at the globe to add any topographical features (i.e. mountains) to their model Earth.
5. Finally, set the paper mache models in a warm, moisture-less location to dry out overnight.
6. The next day, have the students paint their planet using the picture as a guide.

Phases of the Moon

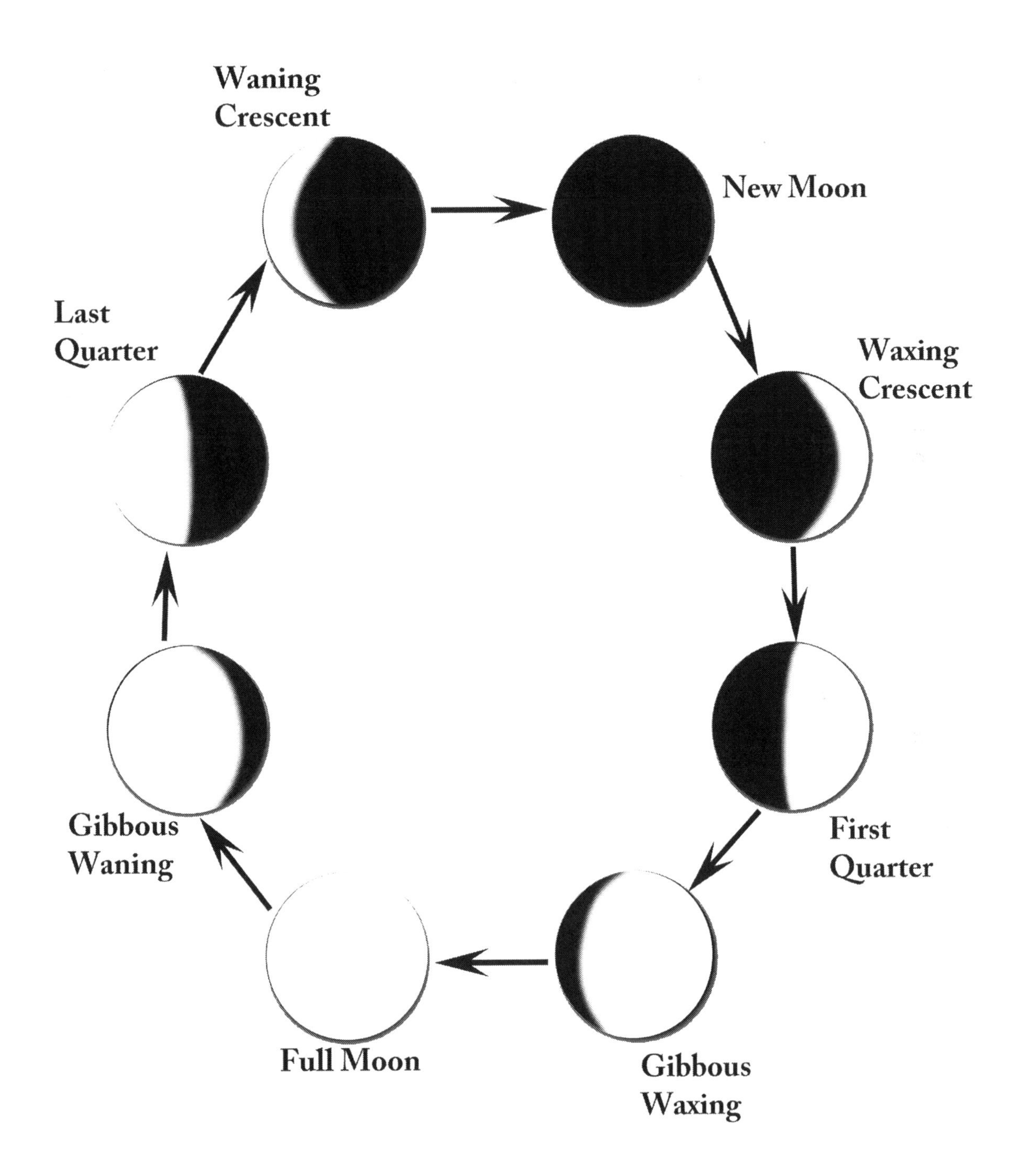

Glossary

Year A Glossary

A

- **Air** – A mixture of gases that form a protective layer around the Earth.
- **Aquatic Biome** – A biome with an abundance of water, including oceans and wetlands.
- **Asteroid** – A rock orbiting the Sun.
- **Atmosphere** – A layer of gases that surrounds a planet.

B

- **Biome** – A very large community of living things, both plants and animals.
- **Bud** – A swelling on a plant stem containing tiny flower part ready to burst into a bloom.

C

- **Cone** – A type of dry fruit produced by a conifer.

D

- **Decibel (dB)** – The unit of loudness.
- **Desert** – The driest biome in the world.
- **Dominant trait** – A characteristic that overrules the lesser seen recessive trait.
- **Drought** – A long period without rain.
- **Dwarf planet** – A celestial body that looks like a planet but does not meet the three requirements to be one.

E

- **Electron** – A negatively charged particle in an atom.
- **Electron shell** – The region around an atom's nucleus in which a certain amount of electrons can reside.

F

- **Flower** – The reproductive parts of a plant.
- **Forest** – A biome characterized by an abundance of vegetation, especially trees.

G

- **Genetic trait** – The characteristics, such as the color of a seed, that are inherited from a parent.
- **Grassland** – A biome with vast grassy fields.
- **Gravity** – The force that pulls objects, such as the Sun and Jupiter, towards other objects.

H

- **Hard water** – Water which contains a lot of dissolved minerals.
- **Hybrid** – A plant that is produced by crossbreeding.

I

- **Interference** – The effect that occurs when two waves meet.
- **Isotope** – An atom that has a different number of neutrons and so has a different mass number from the other atoms of an element.

J

K

L

- **Leaf** – The part of the plant that makes the food for the plant.
- **Lens** – A curved transparent surface that causes light to bend in a particular way.
- **Light** – The electromagnetic waves of energy that make objects visible.
- **Longitudinal wave** – A wave that vibrates in the same direction as it travels.

M

- **Meteor** – A rock that travels through space and burns up when it enters a planet's atmosphere; also known as a shooting star.
- **Mirror** – A shiny surface that reflects nearly all the light that hits it.
- **Molecule** – A substance made up of two or more atoms that are chemically bonded.
- **Moon** – A mini-planet in orbit around another planet.

N

- **Neutron** – A neutral particle in an atom.

O

- **Orbit** – The path of an object in space.

P

- **Planet** – A large ball of rock or gas that travels around a star.
- **Polar biome** – A biome with little vegetation and very cold temperatures.
- **Primary colors** – Colors from which all other colors can be made. The primary colors are red, yellow, and blue.
- **Proton** – A positively charged particle in an atom.

Q

R

- **Recessive trait** – The characteristic overruled by the dominant trait.
- **Reflection** – The change in direction of light rays that occurs when it hits an object and bounces off.
- **Refraction** – The bending of light rays caused by light passing through substances with different densities.
- **Resonate** – To vibrate at the same frequency as something else.
- **Roots** – The part of the plant that anchors the plant firmly to the ground and absorbs water and nutrients.

S

- **Secondary colors** – Colors that can be made by mixing two primary colors. The secondary colors are orange, green, and purple.
- **Seed** – The part of the plant that contains the beginnings of a new plant.
- **Shadow** – A dark area that is formed when an object blocks out light waves.

- **Solar system** – A group of planets and other objects all in orbit around the Sun.
- **Solar wind** – A stream of tiny particles that blow off the Sun and into space.
- **Sound wave** – A mechanical wave that carries sound energy through a medium.
- **Stem** – The part of a plant that holds it upright and supports its leaves and flowers.

T

- **Temperate zone** – Regions that do not experience extremes, so they have warm summers and cool winters.
- **Transverse wave** – A wave that vibrates at right angles to the direction of travel.
- **Tropical zone** – Regions that are typically hot year-round.

U

V

W

X

Y

Z

Review Sheets

Plants Review Sheet

1. A _____________ is the part of the plant that makes the food.

leaf stem flower

2. Circle the name of the process where light energy is turned into food for a plant.

respiration photosynthesis churning

3. **True or False:** Pollen is made in the male parts of the flower and fertilizes the female parts of the flower.

2. Circle all of the things flowers do for the plant.

Produce seeds Attract insects

Are the reproductive part of the plant

5. **True or False:** Seeds contain a baby plant.

6. Fruits help to _______________ seeds.

protect disperse protect & disperse

7. Cones are ________________ produced by conifers.

dry fruits   needles   leaves

8. **True or False:** Spores are tiny copies of seedless plants.

9. Circle all the things the stem does for the plant.

Holds up flowers   Supports the plant

Transports food & water   Makes food

10. Label the following items on the plant cell below - cell wall, nucleus, and chloroplasts.

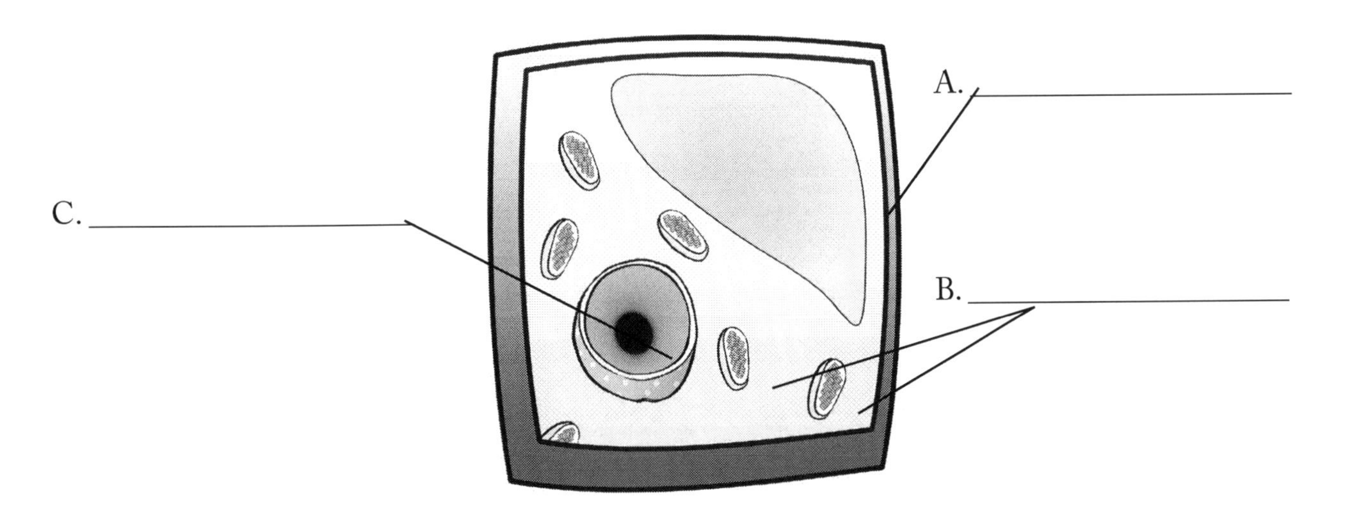

11. Match the type of root to what it does.

fibrous root	grows down
taproot	grows out

12. **True or False:** Roots suck up water and nutrients from the soil.

Mendel Review Sheet

1. Match the following words to their definitions.

_____ Dominant trait

_____ Genetic trait

_____ Hybrid

_____ Recessive trait

A. A plant that is produced by crossbreeding.

B. The characteristic overruled by the dominant trait.

C. A characteristic that overrules the lesser seen recessive trait.

D. The characteristics, such as the color of a seed, that are inherited from a parent.

2. **True or False:** Universal laws explain why some things will always act in the same way, even in different settings.

3. **True or False:** Gregor Mendel believed that plants and animals did not pass traits down from parents to their children.

4. **True or False:** Gregor Mendel just let his plants grow and did not control the breeding of his pea plants at all.

5. **True or False:** Gregor Mendel discovered that some traits are dominant and some are recessive.

6. Why do you think someone should learn about Gregor Mendel?

____

____

____

____

____

Major Biomes Review Sheet

1. The Arctic is near the (North / South) Pole.

2. **True or False:** The polar biome is the hottest boime on Earth.

3. Forests are (all / not all) the same.

4. Rainforests are some of the ______________ places on Earth.

    driest   wettest

5. Most deserts are extremely (hot / cool) during the day and (hot / cool) during the night.

6. African grasslands are called _________________.

    savannas   grasslands   prairies

7. **True or False:** During a drought, the conditions are typically dry and dusty.

8. **True or False:** Wetlands can have only fresh water.

9. Match the ocean to where it is located on the map.

_____ Pacific Ocean

_____ Atlantic Ocean

_____ Indian Ocean

_____ Southern Ocean

_____ Arctic Ocean

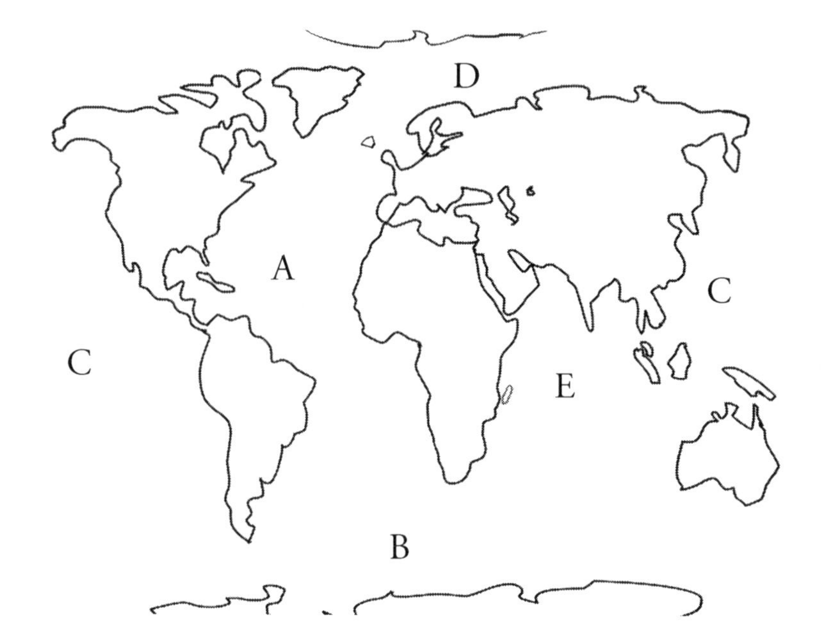

Solar System Review Sheet

1. Our solar system is located in the __________ galaxy.

    C590 Milky Way Kuiper

2. **True or False:** The solar system is a group of planets and other objects that orbit around the Sun.

3. The Sun is a _______________.

    planet rock star

4. The Sun is really a huge ball of ____________.

    gas orange juice liquid

5. Mercury is the ________ planet from the Sun.

    2nd 4th 1st

6. **True or False:** A planet is a huge ball of rock or gas that does not travel.

7. The atmosphere on Venus traps heat from the ___________.

    Mercury sun outer space

8. **True or False:** Atmosphere is a layer or layers of gas that surround a planet.

9. Most of Earth's surface is covered with ________________.

rock water grass

10. A moon is a mini-planet that orbits around ___________________.

another planet a star the Sun

11. Mars is a lot ___________ than Earth.

hotter colder

12. **True or False:** The temperature of a shaded surface will be cooler.

13. Jupiter is the _____________ planet in the solar system.

smallest largest

14. **True or False:** Jupiter's red spot is not moving in any way.

15. Saturn's rings are made up of ________________________.

rocks and ice gas and water fire and air

16. Saturn is a very ___________ planet.

heavy average light

17. Uranus is mostly made up of _________.

rock iron gas

18. **True or False:** An orbit is the path of an object in space.

19. **True or False:** Neptune has the worst storms of the solar system.

20. Gravity is the force that pulls objects _________________ other objects.

towards away from

21. **True or False:** Pluto is very far from the Sun.

22. Pluto, Eris, and Ceres are all ________________.

dwarf planets planets stars

23. An asteroid is a _____________ orbiting the Sun.

planet rock star

24. **True or False:** Comets are large balls of cotton and sand.

25. Meteors are also called ______________________.

really cool shooting stars planets

Atoms and Molecules Review Sheet

1.Match the following subatomic particles with their charge.

Proton	Neutral
Electron	Negative
Neutron	Positive

2. **True or False:** An isotope is an atom that has a different number of neutrons.

3. Fill in the blanks with the number of electrons found in the shell.

4. **True or False:** A molecule is always made up of only one element.

5. Circle the two main gases that are found in air.

oxygen argon nitrogen chlorine

6. Animals take in (oxygen / carbon dioxide) and release (oxygen / carbon dioxide). Plants take in (oxygen / carbon dioxide) and release (oxygen / carbon dioxide).

7. Hard water has (more / less) dissolved minerals. Soft water has (more / less) dissolved minerals.

8. **True or False:** Surface tension is caused by the attraction of the molecules found in a liquid.

Light Review Sheet

1. **True or False:** Light is a form of energy made from an electromagnetic wave.

2. When light passes through a(n) _______________ object, it creates a shadow.

   transparent   translucent   opaque

3. What are the seven colors that make up visible light?

   R ______________________________

   O ______________________________

   Y ______________________________

   G ______________________________

   B ______________________________

   I ______________________________

   V ______________________________

4. **True or False:** Different colors of light have the same wavelength.

5. What happens when light is reflected?

____

____

____

6. What happens when light is refracted?

____

____

____

7. **True or False:** A mirror is a curved transparent surface that causes light to bend in a particular way.

8. **True or False:** A lens is a shiny surface that reflects nearly all of the light that hits it.

Sound Review Sheet

1. **True or False:** Sound is a mechanical wave that carries energy through a medium.

2. **True or False:** An echo of a sound wave cannot be use to determine the position of an object.

3. **True or False:** Mechanical waves, such as water waves and sound waves, cause vibrations in solids, liquids, or gases.

4. Match the type of wave with its description.

____ Longitudinal Wave	A. A wave that vibrates at right angles to the direction of travel.
____ Transverse Wave	B. A wave that vibrates in the same direction as it travels.

1. **True or False:** A wave never changes speed, direction, or shape when it passes into a different medium (substance).

2. ______________________ is the effect that happens when two waves meet.

7. Match the instrument group with how it works.

____ Stringed Instruments

____ Wind Instruments

____ Percussion Instruments

A. Sound is produced by vibrating strings and then is made fuller and louder by resonating in a soundbox in the instrument.

B. Sound is produced by beating, scraping or shaking the outside and then is amplified by the hollow shape of the instrument.

C. Sound is produced by vibrating a mouthpiece and then is amplified as it travels through the tubes of the instrument.

Made in the USA
Columbia, SC
15 June 2024

36663210R00076